Plastic Blue Marble

Catalyst Book Series

01 Plastic Blue Marble—Catalyst: Amanda Boetzkes

02 Phono-Fictions and Other Felt Thoughts—Catalyst: Eldritch Priest

Series editors

Amanda Boetzkes
David Cecchetto
Ted Hiebert

Plastic Blue Marble

Catalyst: Amanda Boetzkes

Edited by Ted Hiebert

Contributors

Shannon Bell
David Cecchetto
Mark A. Cheetham
Ted Hiebert
Doug Jarvis
Serena Kataoka
Anne Pasek
Andrew Pendakis
Maria Whiteman

Noxious Sector Press
Victoria • Seattle

Editor: Ted Hiebert
Contributors: Shannon Bell, David Cecchetto, Mark A. Cheetham, Ted Hiebert, Doug Jarvis, Serena Kataoka, Anne Pasek, Andrew Pendakis, Maria Whiteman.

Cover image: Maria Whiteman, *Astro Was Here*, 2015-16. Photograph.

Noxious Sector Press
noxious@noxioussector.net
http://www.noxioussector.net

Library and Archives Canada Cataloguing in Publication

Plastic blue marble : catalyst, Amanda Boetzkes / Ted Hiebert, editor ; with essays by Shannon Bell [and eight others].

(Catalyst book series ; volume 01)
Essays and illustrations.
ISBN 978-0-9936058-2-6 (paperback)

1. Boetzkes, Amanda--Criticism and interpretation. 2. Art and science. 3. Technology and the arts. I. Hiebert, Ted, 1973-, editor II. Bell, Shannon, 1955-, author III. Series: Catalyst book series ; v. 01

N72.S3.P53 2016 701'.05 C2016-904814-4

Catalyst

The twenty-first century is a time of prodigious creative and intellectual experimentation, with many thinkers, artists, and makers engaging in a range of practices that are foundationally speculative yet nevertheless transformative. The Catalyst book series aims to represent this space of possibility by coupling theorists and artists in ways that galvanize logics, spaces, politics, and practices that are not yet mapped … and perhaps never can be.

Catalysis instigates processual differentiations over a space of exchange; it is eventful, unpredictable, and generative. To chart a catalyst is to bring attention to the critical and creative processes that reveal hidden perspectives upon the event of their becoming. Thus, contributors to the Catalyst books think *alongside* the catalyst, edging and forging implications, connections, atmospheres and weirdnesses. The essays do not review or critique the catalyst's work but rather sound points of contact in pursuit of resonances, enacting gestures of performative solidarity through intellectual and creative engagement.

Catalyst books build speculative communities, inviting a wide range of perspectives into conversations about shared artistic, political, and intellectual values while privileging the unique, distinct and personal insights that characterize any single voice of engagement. Each volume in the series provides an in-depth look at an active thinker or artist—seeking after the full relevance of their work. The series focuses in particular on voices that have not already been widely featured but who have unique and relevant perspectives to share on questions of art, theory and culture.

Contents

Catalyst: Amanda Boetzkes 1

Introduction 3
Ted Hiebert

1 Eight Thoughts on Scale, the City and *The Ethics of Earth Art* 15
Anne Pasek

2 收破烂的人 (*Shou Polan de Ren*): On the Historical Materialism of Scavenging 27
Andrew Pendakis

3 *The Ethics of Earth Art* as Catalytic Theory Converter 41
Mark A. Cheetham

4 Astro Was Here 59
Maria Whiteman

5 Ecologies of (Imaginary) Friendship 79
Ted Hiebert

6 Feeling with Your Eyes: Petroculture and *Ice Follies* 97
Serena Kataoka

7 Facing the Elemental: Amanda Boetzkes, Emmanuel Levinas and *Shooting Theory* 125
Shannon Bell

8 An( )Alibic Aural Tetrad: A Fourfold Structure of Ecologicity 141
David Cecchetto

9 Belly Brain: An Instrument of Ecological Perception 159
Doug Jarvis

Contributors 183

Catalyst: Amanda Boetzkes

Amanda Boetzkes is a theorist of contemporary art and visual culture with a particular focus on the intersection of creative practices with the biological sciences, particularly ecology and neurology. Her first book, *The Ethics of Earth Art* (University of Minnesota Press, 2010), considers the ways in which contemporary art developed an ethical concern for the earth and a corresponding aesthetic sensibility borne of elemental thinking.

Her co-edited collection, *Heidegger and the Work of Art History* (Routledge Press, 2014), explores the impact and future possibilities of Heidegger's philosophy for art history and visual culture, and considers the ontological and ethical implications of our encounters with works of art; the visual techniques that form worlds; how to think about "things" beyond human-centered relationships; the moods, dispositions, and politics of art's history; and the terms by which to rethink aesthetic judgment and the interpretation of the visible world, from the early modern period to the present day.

She is completing a manuscript entitled *Contemporary Art and the Drive to Waste* which examines the interplay between the aesthetics of contemporary art, global systems of energy-use, and the life cycle of garbage. Her upcoming book project, *Ecologicity: Vision and Art for a World to Come* analyzes the aesthetic and perceptual dimensions of imagining the ecological condition. Boetzkes has published in the journals *Postmodern Culture*; *Art History; Reconstruction: Studies in Contemporary Culture*; RACAR; *Antennae: The Journal of Nature and Visual Culture*; *Eflux;* and *nonsite.org*, and has contributed chapters to numerous books and catalogs, including *Marxism and Energy* (MCM', 2016); *Fueling Culture: 101 Words for Energy and Environment* (Fordham University Press, 2016); *The Edinburgh Companion for Animal Studies* (Edinburgh University Press, 2017); W*est of Center: Art and the Countercultural Experiment in America, 1965-77* (University of Minnesota Press, 2011); and *Art History: Contemporary Perspectives on Method* (Wiley-Blackwell, 2010).

Boetzkes is Associate Professor of Contemporary Art History and Theory at the University of Guelph, and lives in Toronto (Canada). https://amandaboetzkes.com

Introduction

Ted Hiebert

Among the best catalysts for creative and speculative thinking are scholars who work across disciplines, merging discourses in new and thoughtful ways or rubbing the edges of different perspectives against one another. This is so noteworthy an activity precisely because there are no established methods to follow; with no dominant discourse as a guide, a thinker is called upon to invest in a uniquely constituted version of the trajectories he or she explores. Such thinkers produce ideas that are constitutive rather than demonstrative, dialogic rather than authoritative, and conscientious of the relationships they build with other thinkers of similar ilk. That is, thinkers whose ideas catalyze others are often themselves thinkers of catalysis, targeting the mixtures, liminalities, hybridities and alchemical potentiality of thought, as Antonin Artaud once put it, as a form of social and intellectual incantation.

There is no better example of such a catalytic thinker than Amanda Boetzkes, whose work spans areas ranging from contemporary art, theory and criticism to earth art, ecology and ethics—from questions of waste and the archaeology of garbage to questions of embodiment, subjectivity and theories of consciousness. Throughout her work, Boetzkes also shows a deep sensitivity to the question of technology, whether seen as new media or old, genetic, neural or visual. It is not simply that Boetzkes's work touches on all of these areas of study, but that she also relentlessly entangles them with her own thoughts, as attentive to gaps, interstices and intellectual openings as she is to the implications of her analyses on other related and like-minded fields.

In a broad way one might say that this arc of ideological entanglement in Boetzkes's work resonates with particular strength at the intersections of art, history, ecology and the study of technology. It is this particular combination of foci that make her an important catalyst for this volume—*Plastic Blue Marble*—which builds tangential resonance around several of Boetzkes's principal arguments. One might even situate certain key thoughts at these intersections, not in an attempt to exhaust the wide range of questions that Boetzkes engages but rather to accentuate the catalytic potential of her work, seen here as a series of three disciplinary cross-overs.

At the intersection of art history and ecology lies the realm of earth art, what is often called "land art" except that Boetzkes refuses to desegregate land and ecology such that in the study of earth art lie the terminologies for how we might re-imagine our relationships to the planet itself. This is also the central theme of Boetzkes's first book: *The Ethics of Earth Art*. It is here that Boetzkes establishes her vision for a unique form of ethical and ecological relationality, engaging the complexity of artistic production for its ability to represent the limits and the paradoxes of ethics and ecology, seeking ways to build authentic interactions with the world around us while at the same time insisting on accountability for the ecological damage for which we ourselves are responsible. Such a perspective could not be thought by an artist, an historian, an ethicist or even an ecologist alone, but requires a uniquely interdisciplinary sensitivity to each of these domains in order to weave together a mandate for conscientious personal engagement and (also) a new proposition for ways to consider the social, aesthetic and ecological implications of artists working with and on the earth.

At the intersection of technological culture and art history lies the question of the medium (or the tool), a concept that Boetzkes activates with particular force by bringing together the often separate discourses of artistic and technological media. It is not simply that Boetzkes attends to the implications of media theory on the understanding of art (McLuhan's famous "the medium is the message," for instance, as an utterance that might be applied to painting or sculpture no less than to McLuhan's favorite examples of television or radio), she also does the inverse. That is, one of the defining features of Boetzkes's work is that she understands deeply the value of an art historical imagination that can be applied to the present day questions of technology and culture. This is one of the central through lines of her co-edited book (with Aron Vinegar) *Heidegger and the Work of Art History* which begins with the attempt to "rethink aesthetic judgment and the interpretation of the visible world" but ends up going much further, inviting other scholars and artists to broaden the exploration of Heidegger's work on the question of technology and in so doing establishing a forum for the collaborative reconsideration of the relationships between technology and art historical practice.

At the intersection of ecology and technology Boetzkes's theorizations push the possibilities of technological reconsideration even further, examining questions of social and political intervention and the ways in which art historical understanding has begun to recast itself as a social and ecological practice. It is here where Boetzkes insists on an ecological understanding of art history and technology alike, as modes of engagement that inform our understanding of the age of the Anthropocene. For this is a time in which human intervention on a global scale threatens to irretrievably alter (if it hasn't already) the direction of planetary evolution, and as such creative, speculative and immanently self-reflexive modes of thought are urgently required for understanding the complexity of the ecological present. These might be seen as technologies of and for the process of understanding our relationship to the world differently. They are also topics of recent essays and upcoming books (*Contemporary Art and the Drive to Waste* and *Ecologicity: Vision and Art for a World to Come*) in which Boetzkes seeks to understand the ways in which contemporary art and the logic of global capital share an ideology grounded in economies of plastic, garbage and waste.

For Boetzkes, living as we are in the aftermath of global climate change, it is not simply objects and materials that have become plastic but consciousness itself, and our idea of the planet we inhabit must change as a result. Our earth is no longer the solitary blue marble pictured from outer space by the Apollo 17 crew in 1972. Now, scholars such as Amanda Boetzkes imagine it as a *plastic* blue marble, mediated as it is through the paradoxes of intersectional and elemental thinking, anthropogenic change, and the ongoing project of visualizing the futures we are building together. Neither futile nor redemptive, life on the plastic blue marble is instead tangled and complex, requiring not just speculative forms of thought but emergent minds that are as attuned to the relationships they create as they are to the serious need for creative and collaborative thinking. Amanda Boetzkes is one of the thinkers who can provide such an interesting starting point for building these kinds of relationships because her work not only serves as a brilliant catalyst for others but is itself deeply aware of its own status as catalysed in turn by the various relationships and questions she explores.

Plastic Blue Marble

This volume brings together a series of different perspectives, each in their own way catalyzed by the work of Amanda Boetzkes, though not necessarily bound to her perspective, faithful to notes of resonance but deeply individual in their speculative concerns. From visions of catastrophe to poetic journeys through the urban, social and artistic imagination, the contributions in this volume attune to Boetzkes's method as much as to her thoughts in order to explore and propose a constellation of strategies for what it might mean to engage with the questions of art, ecology and technology while living on a *Plastic Blue Marble*.

The volume begins with Anne Pasek's "Eight Thoughts on Scale, the City and *The Ethics of Earth Art*" which is a compelling meditation on key ideas from Boetzkes's book, highlighting a general context of ecological and political engagement, and the distinctiveness of the author's own encounters with the city. Opening with the image of a photogram written in smog—a literal moment of inscription and a metaphor for Pasek's larger theorization—the essay insists on an incommensurability of the experiential moment as a horizon to which receptivity must aspire. This is true, particularly, if one is to maintain a relationship with what Pasek eloquently call the "inscriptive force of the city." At stake in this inscriptive force of the city is the insistence that an ethics of ecology depends not just on perspective, but also on the (sometimes paradoxical) interplay of ideas and engagement (or framing and perception). Theorized as a series of "mutually unwilling collaborations" between the city and its residents, Pasek stresses the importance of perceptive participation, thinking about urban context for the relationships it builds—even when they are not always friendly.

Building on Pasek's notion of "inscriptive forces," Andrew Pendakis's essay—"收破烂的人 (*Shou Polan de Ren*): On the Historical Materialism of Scavenging"—meditates on the *shou polan de ren,* residents of urban Chinese cities who build mounds of garbage on top of bicycles and carts. These "masters of an art and science of bundles," as Pendakis calls them, are not just scavengers but collectors and consolidators. Entering into the social and political complexities of building culture out of garbage, Pendakis theorizes the temporal present as already a space of ruin, proposing the act of scavenging as the future of productive labor. In his analysis of the *shou polan de ren*, Pendakis finds a parallel discourse to the art historical—thereby also providing an alternative to the tendency of many philosophers to romanticize the artist as the trend setter for future thinking. For Pendakis, it is not the artist that holds the secret to future living but the *shou polan de ren*, building their mobile towers out of scraps on the street. The *shou polan de ren* live within (or after) the ecological catastrophe, embodying a social proximity that reinforces our (and their) role in the ecosystem of urban excess.

The notion of the catastrophe is central also to Mark Cheetham's chapter—"*The Ethics of Earth Art* as Catalytic Theory Converter"—which recasts Theodor Adorno's prohibition on poetry in the aftermath of WWII for the ways it can help understand a culture of ecological disaster. In particular, the essay creates a dialogue between the ethical complexity of ecological living and the representation of nature—using examples such as Hans Haacke's attempts to "articulate something natural" and Stewart Hall's observation that articulation is an act of holding together potentially contradictory forms. In so doing Cheetham proposes a recapitulation of the sublime as a marker of art with potential to implicate a viewer as participant and not merely as an observer. Such a position reinforces something central to the possibility of encountering the sublime in representation—that a painting is also an experience waiting to happen. The result is a critique of the conventional interpretation of representation as static, insisting instead on the dynamic life of the image. In this, Cheetham implicates the limits of critical human perception, eloquently framed as "intimating a limit of human forms," a bringing-together of the seemingly antithetical forces of representation and ethical encounter.

The idea that representations can harbor an ecological conscience is a central tenet of Maria Whiteman's photo-essay, "Astro Was Here," a series of images of a toy astronaut situated in deserted, broken and ruined landscapes. Each photograph is paired with a quotation that serves to juxtapose the complexity of Boetzkes's ecological conscience with Whiteman's eerie poetics of ruin. In Whiteman's imagination the futurity of the technological image pairs itself with an apocalyptic sublime such as to remind us of the fragility of the planet while accentuating our own states of human alienation. "Astro" in these photos is a metaphor for each of us but more than that as well—a reminder that the future we are making is one we ourselves may no longer be present to witness. It is a decidedly prescient thought, since the consequence of not living in the future is that the stakes of how we imagine it increase. That which cannot be lived (ruined as it will surely be for the future of human survival) can only thus be imagined. And Whiteman's essay places each of us in this imagined future, as tourists of an ecological ruin we have also created.

My essay—"Ecologies of (Imaginary) Friendship"—continues the theme of imagining in ways that resonate with the charges of Cheetham and Whiteman's work, that is, the attempt to imagine both the future we are destroying and its artistic consequences. The essay meditates on an ecological study in which scientists flew drones around a family of black bears, juxtaposing some of the ironies and nuances of the experiment with artworks by Joseph Beuys and Marc Dion. In so doing, the essay argues that the stakes of re-imagining our relationship with the ecological world have never been higher. Taking literally Boetzkes's suggestion that the tools of art criticism have particularly salient value for the understanding of our ecological present, the essay explores what it would mean to imagine science, and indeed the future we are collectively generating, as itself a work of art already under production in the present.

Taking the idea of "life in the present" one step further is Serena Kataoka's chapter, "Feeling with Your Eyes: Petroculture and *Ice Follies*," which begins the difficult project of searching for ways to orient ourselves to the disorienting climate of contemporary social and political existence. Mobilizing Carlos Castenda's proposition that crossing one's eyes is one way to see the world differently, Kataoka meditates on what it would mean to think about contemporary life as literally doubled—caught between the harsh realities of economic competition and social injustice, on one hand, and the desire to invest in one's personal and geographic context, on the other. Using her own context of living in northern Canada as a guide, Kataoka likens the complexity of this doubled gaze—or "fuzzy vision," as she calls it—to a form of embodied and artistic practice. Taking the analysis one step further, Kataoka then turns to a winter art festival in central Canada—*Ice Follies*—as a site from which to begin thinking about affective engagement as a way to "feel with your eyes," a compelling suggestion for including emotional, imaginative and relational points of reference in our understanding of the metrics of life.

The idea of affective resonance is also central to Shannon Bell's contribution—"Facing the Elemental: Amanda Boetzkes, Emmanuel Levinas and *Shooting Theory*"—an exploration of elemental thought in visual and theoretical form. Suggesting that "one ought not to think political theory simply within language," Bell demonstrates how a hybrid form of visual and conceptual engagement is able to open up territory for affective, creative and relational forms of political thought. She does this by highlighting the relational intensities and moments of resonance among thinkers such as Emmanuel Levinas, Simone Weils and Amanda Boetzkes, grounding her thoughts with images of the Alaskan landscape, and in so doing bringing ecological reality to philosophical conceptualizations of the "elemental" and "attention." Framed through the concept of an "*ex post facto* influence"—a way to conceptualize the dialogic destiny of intellectual and artistic engagement—

Bell shares her prior engagements with elemental thinking and the ways in which forms of hybrid, attentive and visual engagement can intensify the experience of elemental and artistic living. In the context of her *Shooting Theory* project—which elaborates key philosophical and political concepts in visual form—Bell suggests tools through which we can begin to build different forms of attention to the relationships between ourselves and the worlds we inhabit.

David Cecchetto's chapter—"An( )Alibic Aural Tetrad: A Fourfold Structure of Ecologicity"—examines the idea of relationality from a slightly more technical (if no less poetic) perspective, exploring the complex nuances of mobile technology and embodied experiences of vertigo, with particular attention to the ways that experiential living, and especially sound, accentuate the generative potential of relationships. Atuning his thought to Boetzkes's theorization of "ecologicity" as a form of self-aware environmental consciousness, Cecchetto holds together three meditations—on listening, on gathering, and on distributed agency—such as to attune to the registers of embodiment and disappearance: the non-saying of saying, sums that are more and less than their parts, and the idea of second-order misdirection taken to the level of autonomic reaction. In this way Cecchetto recasts the act of listening for its resonance with the spirit of ecologicity such as to position listening as an act of questioning rather than simply a process of perception. In many ways, listening seen through Cechetto's constitution, becomes less about hearing in an informatic sense and more about productive disorientation, and the feelings that rise to the surface as autonomic directives at the moment when embodied vertigo sets in.

Taking up the act of listening in a slightly more direct if equally disorienting way, Doug Jarvis's "Belly Brain: An Instrument of Ecological Perception," is a series of creative speculations on communication with non-human entities. Beginning with a discussion of the scientific discovery of neurons in the stomach, Jarvis shares recent manifestations of his artistic practice—including various attempts to record the neural activity of his belly, durational performances designed to re-cast his relationship to his own body, and group activities in online environments. In each case, the projects explore aspects of what it might mean to take the belly seriously as a site of interaction, and indeed conversation. The belly brain, for Jarvis, is both an instance of foreign agency (requiring new communication strategies) and a metaphor for our entanglement with other entities, be they bacterial, virtual or ecological. By creating highly speculative situations and absurdist performances, Jarvis asks not just how we might communicate with entities beyond our frames of (human) understanding, but how we might also learn to listen differently to the sounds and messages such entities might be trying to share with us.

Taken together, the essays in this volume redistribute the currents of Boetzkes's coordination of ecological perception, art theory and history. They discover new terrains of consideration, styles of thinking and creative forms within the terms of the *Plastic Blue Marble*. Moreover, they invite the reader to embark on further entanglements, reflections, articulations and creative acts, in response to the richness of their own catalytic thinking. Life on the plastic blue marble might be complex and, at times, daunting, but what the thinkers in this volume collectively demonstrate is that by attuning to the myriad possibilities of grounded, thoughtful engagement we can indeed reveal new ways to conceptualize, imagine, produce and politicize the possible futures that bind us together. In many ways, such a collection then demonstrates in true form one of the overarching trajectories so key to understanding the work of Amanda Boetzkes—as theorist, scholar and catalyst—namely the importance of attuning ourselves to the relationships we create with the Earth itself, not simply as an object of understanding but as a collaborative condition of artistic, political and speculative existence on this *Plastic Blue Marble.*

1 Eight Thoughts on Scale, the City and *The Ethics of Earth Art*

Anne Pasek

I.

Three months after I move to New York, I notice that the city has left a murky impression on my windowsill. An oily veneer, a topcoat of pollution has formed across the white painted ledge, covering it in sticky soot as equally as the objects I've laid out to rest on top of it. When I remove them for a worried cleaning, I notice the hazy negative space of their irregular forms beneath, small patches of wood spared the smudging touch of the city air. This peculiar photogram, writ in smog rather than in light, strikes me as equally alchemical as photography, although somehow more invasive, more viscously haptic. I wipe the image clean with difficulty, rubbing the particles first into ropey lines of grime before they come free.

This is the first of many such pictures the city draws for me.

II.

This is the window in front of which I've set up my desk. I have Amanda Boetzkes's *The Ethics of Earth Art* in front of me soon after, its pages dog-eared and underlined. I am in the middle of preparations for my comprehensive exams, compiling notes on a conversation that has been unfolding across eco-criticism, literature, art history, and feminist science studies over the past three decades. At times, *The Ethics of Earth Art* seems the odd one out. While many of my readings address the panicked flight from the idea of nature itself, the artists animating this book sometimes seem to be running in the opposite direction.

At the core of Boetzkes's study is a receptive exchange and a subtle inscription between bodies and the elemental forces of the earth. The artists she describes orient us towards an encounter that is both phenomenologically-sensed and constitutive of a marking: on the retina, the skin, the landscape, the page. Dennis Oppenheim's body is darkened by the burning sun; Ana Mendieta's silhouette is imprinted and distributed on the dirt; while Olafur Eliasson's assemblage of mist, light, and the moving spectator strikes a rainbow in the middle of a gallery like fire from a match. As Boetzkes argues, these many generations of earth artists have developed a shared approach to a nature embodied in both perceptually inexhaustible elementals and the traces that they leave on the many surfaces (flesh, stone, paper, water) of their points of contact. These artworks index nature in a way that always seems to disclose the simultaneously rich presence and concomitant epistemological poverty of the image/encounter itself—a perceptual IOU that can never be redeemed, but whose possession itself seems to be the lesson. This kind of receptivity, Boetzkes suggests, is both a material quality of the surfaces of art-making and a practiced ethical stance towards a nature that can never be fully defined, sensed, or possessed.[1]

Most of the other texts on my list, to the contrary, seem to anxiously avoid such encounters, dedicating themselves instead to the intellectual labor of deconstructing old divisions between the human and the natural as a matter of first principles. This isn't to say that Boetzkes's writing is irrevocably tangled in a dualist web, but, rather that in the effort to break ecology out of the pastoral countryside, her fellow ecocritics and philosophers have atomized and scattered nature to the winds to the point that it becomes rather difficult to rehearse the Levinasian drama of a perceiving human subject and an unknowable Other. For example, Bruno Latour and Timothy Morton most famously regard the idea of a categorically-distinct nature as Modernity's original sin and continued inhibition.[2] They prefer to model relations in complex networks and meshworks rather than that of face-to-face encounters. Karen Barad, not too dissimilarly, advocates a retreat from established human/nonhuman ontological differences, seeking to settle these accounts only in the last quantum instance. In her schema of intra-action, perception and (to make an even bolder claim) reality itself are only the local result of coterminous apparatuses that continuously unfold the universe and one another.[3] Similarly, Stacey Alaimo speaks of transcorporeal relations that undercut the distinction between the body and its ecology at a molecular level. Emphasizing the material interchanges between human matter and an increasingly toxic ecology that moves across the human/environment, Alaimo maintains that "'the environment' is not located somewhere out there, but is always the very substance of ourselves."[4] Such approaches to the earth and the body seem at odds with Boetzkes's emphasis on receptivity, insisting instead on the quotidian and the complete movement of environmental forces—but not necessarily sensations—on and across the skin. As such, within contemporary thinking about the environment, it is increasingly difficult to imagine the stage for Boetzkes's ethical gestures: there is an insistent refusal to maintain the distinct cast of characters (a perceiving human subject and an exterior nature) necessary for her artists' receptive images, even as their distinctions are quickly muddied through the act of inscription. At this flattened scale, perceptual limitations often become foremost a matter of political redress rather than a foundation for ethical encounters.

In this turn, I wonder whether or not something of great value has been lost.

III.

There is still the matter of the greasy image on my windowsill—its unsettling, uncanny presence. I worry about those marks, the haze of the city that settles around my desk and presumably also on and around me. I think of the leather of my skin, how it would yield to a charcoal mark, a tattoo needle, the spray paint of industry; I think of the slow tinting of phlegm and teeth through the city's grit; I think of New York's many mutually unwilling collaborations with its residents, the pictures they've made together, erased, and regarded. For all of the new materialisms' many invocations of an always already undone distinction between the body and the environment, is that really how the world is lived, is felt?

I don't want to be a New York assemblage assembled inside out—materially, politically, visually adrift in a horizonless sea of matter and its mattering. Nor, for that matter, am I wholly convinced that this is so, regardless of how convincing I continually find the scientific and rhetorical arguments to this end. Instead, my experiences of the city's often overpowering presence seem at times more akin to that of an indexical encounter than a quiet congress of ontologically-networked parts. I often find myself somewhere in between flat ontic claims of what strictly is and the romantic traditions of what could, with poetics, come to be.

It strikes me that new materialism lacks a theory of scale and of subjectivity. As a corporeal subject, I am all too attuned to the ends of my fingertips, my enduring and highly personalized sense of time, and the shape of the world from the space of my eyes, although I may frequently extend and borrow from other objects, forces, and people. This is a dualism not of militant mind/body divisions, but of an elastic yet inescapable sensory gap between my body and its external prostheses, cohesions, and differences in the world.[5] Like Merleau-Ponty, I can touch myself touching, yet I am lost when I try to conceive of the unctuous particles dispersed along my window touching in the same way, across the same span of experience. The grime's disparate scale and framing along white painted wood ensures that I encounter it as distinctly alien, despite the fact that it is doubtlessly already a part of me in some small way: borne from the fuel I use, the commodities I consume, and long since embedded in my hair, lungs, and gut. Its animacies and provenance, rich and doubtlessly there to be studied, are not fully disclosed to me from the window. Instead, this is a moment of encounter in which divisions are drawn, even if just a moment, so that marks can be read and interpreted.

IV.

In describing the inscriptive force of the city I find art to be a very useful metaphor. I feel like rag paper, circulating in the city's streets collecting the exhaust of tailpipes, the downpour of pollen, the perspiration of cigarettes, incense, sweat, smog, and heat. I am a wire form, slowly beaten into a new shape as I strain my legs along the pavement, twist with the rhythm of the train's lurching step, and bend over my books, squinting. Muscles wax and wane, skin pales and shadows, and tendons grow variably brittle or plastic as I navigate and inhabit the city. Consciously or not, I receive its marks and inscriptions. I am, rarely with any intentionality, an urban earth artwork that indexes the elementals of the city's environment along the envelope of my body and the filaments of its interior. The challenge I might take from Boetkzes's formulation is to become a more perceptive participant of this exchange. If I cannot escape the habit of perceiving myself as a distinct subject acting in a world full of Others (in form, in mind, in scale) then the ethical stance to follow this parallel would be to cultivate an awareness of my body as the object of their inscriptive actions in turn.

In this way, it seems that the city and I rub against each other in the dark, learning about one another by interpreting the smudges left on our skin—be it a crime scene, lover's bite, or an augury. Learning to read and monitor this language is doubtlessly also an important survival tactic: a mode of monitoring slow acts of violence and triumphs, or of forging solidarities between those shaped into similar forms and recognizing inequities in the differences. It might also be the grounds on which to found a new realism; what was an ethical acknowledgement of a lack of dominion and mastery within earth art is more blatantly the case in relation to a city like New York and the spirits of capitalism and ruin that run through it. Attending to how we are scaled in what we find around us could be a good first step.

V.

I've seen New York City from many angles—above, below, over, and through—but I wouldn't say that I know it or consider it a friend.

In writing these words I can't help but personify it, attributing to it the activity of the passive voice of my bodily narration. There is of course a rich history of extending such courtesies, of thinking of the city in terms of a body, composed of arteries of circulation, waste pipes, nourishment, growth. and decay. A sick city, a city in glut, a city cleansed, a city that never sleeps. Flying low over Queens and peering shyly out of an airplane window at dusk, I see thousands of lights moving, pulsing, obscuring one another and then returning in force as cars, trains, and people make their way across the city to the next place they need to be that evening. It is almost impossible not to imagine this glowing crowded grid as blood flowing through a vein, full of cells looking to nourish or devour the subjects of their encounter.

From this vantage point, my windowsill image and my rag paper body are inverted, for it is also the city that is variously beaten into shape by the subjects of its bearing, the many steps the crowd takes along its pavement and parks, the forms we build, break, and leave broken. It is a vast photograph, exposed under the shadows of millions of stirring bodies, or a pearl built up from ongoing irritations in our shared mucus. The city refers to its people in parts: fragments of ambitions, compassion, habits, ill-intents, and communities coarsely externalized into buildings and dust. We act on it just as it acts on us, though neither route is equally felt or directly traveled in the passage through the ontological gauntlet of shifting scales.

VI.

This gap—the jump from one human body to one urban form and back again—is a site of profound disorientation. It is a space in which solidarities often fail and can never be easily presumed. Uneven actions, translated and migrated, come to form something categorically different and internally animate. These elementals of the city—its pollution, noise, and its sublime moments of distress and beauty—are authored collectively though we do not often recognize ourselves in the marks that we come to bear from these forces in turn.

There is a kind of mutually disparate and unfelt receptivity between entities divided by these kinds of scales. I buy plane tickets and till earth, occupy park benches, scrawl phrases on walls, and participate in the exchange of goods and ideas in and out of the city's borders. Similarly, the city marks down and it marks up, writing on components that are nested inside it (down onto my windowsill, my neighbors' lungs, the birds that cross its waters, or the fish of its harbors—up into larger biospheres and environments, painting the Hudson a different hue, directing global flows of capital, or thickening the air over the Atlantic with carbon and sulfur). These activities are connected in space and in agential capacities, though rarely consciously so. Instead we refer to one another across bodies so disparate that perception and accountability are most often impossible.

Techniques for material inscription and bodily perception are required to make these relations more humanly sensible (as encounters with elementals, Others, or a kind of nature) while still acknowledging that these relations can never be wholly sensed. This strained possibility for an ethics across the accumulation and transformation of marks made and received is what is most acutely needed in this moment of ecological peril.

VII.

There is already a set of gestures to help chart these relations of incommensurability and inscriptive reference, one again that can be borrowed and adapted from the practices of earth art. The most notable parallel is that of Robert Smithson's site/nonsite relations: works of art that, like the slow creep of pollution or the accounting of a carbon footprint, organize relations across disparate space and frames.[6] In pieces such as Smithson's *A Nonsite, Pine Barrens, New Jersey* (1968), sand from one locale (the distant plains) is moved to another (the James Cohan Gallery in New York City) and in so doing is made into a "nonsite" that is present in a form that never refers solely to itself but always directs its viewers onwards to a disparately sensed elsewhere.

Site/nonsite relations are often theorized in terms of their spatial logics, but this concept proves useful to the task of thinking across scales as well. As a part from the whole, the nonsite is composed of the same stuff of the site, though in its minutia it shifts in ontological weight and agency. One requires a kind of metaphorical thinking to address this absent presence. The nonsite cannot do what the site can do, yet it is this property of difference that allows the site to be newly communicated through the partiality of its trace. As Boetzkes suggests, this problem of representation—the inability to ever locate the artwork completely in a given moment or in the sum of its perceptions—invites an ethical acknowledgement of our situatedness in space and in size.[7] A sample of sand can communicate something of an incommensurable vastness if its relation to a distant site is maintained and made sensible to the scale of the human body.

Much of the success of this gesture, it seems, relies on the joint efforts of skillful framing and receptive spectatorship. This is certainly the case in the study of the city's traces; just as Smithson pulled earth from a faraway site into the gallery in order to make it newly sensible within the very impossibility of knowing or locating it completely, something of the elementals of New York City were deposited here before me, pointing outwards in every direction to a site that cannot otherwise be known or wholly cataloged.

VIII.

Art is a frame (whether skillfully authored by an artist or rudely imposed by the present author) in which a little breathing room can be found within the vertigo of pancake flat ontologies, the cacophony of cities, or the urgency of global crises. It is a practice, perhaps even a quotidian one, that can provide distance and perspective with which scales can be measured or productively skewed. It is a domain of evident artifice in which thought is less beholden to the impatient demands of what is or what is acutely needed, but is instead given over to the craft of what could be sensible or what we contend to matter. It is, in short, a space in which we might still find the alterity of Others or a categorically distinct Nature, even if this mirage only holds for the duration of our glance.

As Boetzkes reminds us, art provides not just a metaphor to track these transformations, but a model for an ethics to bring to bear on the complex terrains of the earth's many modes of receptivity. It is at this moment of pending environmental catastrophe that we need these ethics all the more. The point is not to recognize ourselves equally and easily in the face of anthropogenic climate change, nor to rehearse possible strategies of making this problem of elemental scale somehow newly visible and knowable in its entirety—as if this were a precondition for action. Our pressing problems might instead demand different epistemological tactics and subtle strategies for action predicated upon our ability to better interpret the marks of nonsites and elemental inscriptions on our bodies, and to become more skillful mark-makers in turn.

Notes

1 Amanda Boetzkes, *The Ethics of Earth Art* (Minneapolis: University of Minnesota Press, 2010), 21.
2 Bruno Latour, *We Have Never Been Modern* (Cambridge: Harvard University Press, 1993); Timothy Morton, *Ecology without Nature: Rethinking Environmental Aesthetics* (Cambridge,: Harvard University Press, 2007).
3 Karen Barad, *Meeting the Universe Halfway: Quantum Physics*

and the Entanglement of Matter and Meaning (Durham: Duke University Press, 2007).

4 Stacy Alaimo, *Bodily Natures: Science, Environment, and the Material Self* (Bloomington and Indianapolis: Indiana University Press, 2010), 4.

5 Donna Jeanne Haraway, "Situated Knowledges: The Science Question in Feminism and the Privilege of Partial Perspective," in *Simians, Cyborgs, and Women: The Reinvention of Nature* (New York: Routledge, 1991), 183–201.

6 Boetzkes, *The Ethics of Earth Art*, 69.

7 Ibid., 145.

2 收破烂的人 (*Shou Polan de Ren*) On the Historical Materialism of Scavenging

Andrew Pendakis

Though rare sights in the West, "scrap collectors" (*shou polan de ren* 收破烂的人) in China are basic constituent units of the contemporary urban streetscape. These workers—often migrants from rural areas who lack the official urban status needed to access government services (or *hukou* 户口)—are some of the most precariously employed in China. On carts pedaled by foot or driven by small motors, *shou polan de ren* scavenge scrap material from factories or the fringes of construction sites and sell it in bulk to local processing plants. Often characterized by compelling chromatic or textural configurations and by scales that appear to momentarily suspend consistent naturalism, these assemblages are not simply modes of transportation but mobile visual events—"floats," as it were—that punctuate urban space with brief shards of improbability and strangeness.

These objects are remarkable, however, not primarily because they are novel or different; variation *per se* is less interesting than it seems and is in many ways the very mechanism by which a system grounded in the capture of attention procures docility and consent. The canonical surrealist surprise—a bullet fired blindly into the crowd—is, *vis à vis* the social whole itself, an example of what Aristotle categorized as "accidental" (positional), rather than "substantial" (generative) change: things move about, but the space in which they do so stays the same.[1] What makes the *shou polan de ren* interesting is not only the way they lead us to "see anew" (surely among the worst of today's theoretical clichés) but the way their labor operates politico-aesthetically on a determinate field of relations—namely a capitalist society organized around the interminable production of new commodities. If—as Adorno and Benjamin taught us—*every object is a pedagogy,* what are the truths diffused by this roaming statuary of trash, this art and science of the poor practiced in full view of the city?

1

In societies based on the exchange and circulation of commodities, the latter are experienced as empiricist particulars rather than as members of a set or class. The object exists as an orphaned "this" or "that," its kinship determined abstractly by the heuristic of type (*a* car, *a* cup), or via the reifying short-hand of brand. A glimpse of the properties shared by the object and others of its kind—its concrete universality, as it were—takes place only in the antechamber of exchange. Merchants have always had to manage the tension between a display that hustles the gaze with pure scale (the primitive splendor of the pile or pyramid) and the need to foreground, despite it all, the object's rarity, its difference from junk. Even before exchange has taken place, desire suppresses the indifference of the commodity—the sameness and exchangeability that effectively defines it—by filling the gap between wanting and owning with the lure of distinction. In such moments, desire is a bad nominalism: it sees nothing but tempting singularities.

After the act of exchange has taken place possession definitively divides the object from the last trace of its own universality: it enters the gated community of use. Locke famously conceived of the difference between the commons and property as mediated by a decisive act of

Andrew Pendakis, *Shanghai*, *China*, 2013.
Color photograph.

labor: the latter introduced into the undifferentiated *hyle* of nature the *morph* of something determinate and owned. A high wall encircles the private theatre of consumption. What matters in this is that reification is not merely the object separated from its process of production, or from social relations, but from its own numericity in which that process and its scale can be discerned in the form of a trace. This conjuncture was powerfully illustrated by the story of a recent factory worker in Guangzhou.[2] The man chose to propose to his girlfriend on the street in a heart he had constructed entirely out of stacked iPhones (99 in total). The phones represented 17 years of salary and an unknown number of concrete labor hours. A gesture he had hoped would function as an irresistible synthesis of premodern virtue and post-capitalist allure completely backfired. His lover's rejection rendered explicit the failure already there at the core of his plan: *an aggregate of commodities always weighs less than the sum of its parts.*

In the time between the scavenged object's first and second death—in other words, between the moment of consumption and that of reclamation—it enters into a state of rare visibility. In this phase, that of *public transit*, the object is returned to the jurisdiction of the universal and in the process is degraded. This is, on one hand, the humiliation added to the object for not being *sui generis*. On the other, the shame here is attached to *simply having been used in the first place.* It is no coincidence that our culture's misogyny translates rape into the idiom of contaminated use. To be *used up* is the ultimate form of social opprobrium. This shame, of course, is nothing more than the value subtracted from fetish by use: it is the disjunction between the unconsummated allure of the commodity and its consumed shabbiness.

No commodity avoids this fate: even the unused become embarrassingly old. Today, the last vestiges of Platonic idealism are not to be found in everyday consciousness nor in the political systems that govern us (both of which run on pure liberal skepticism), but in that seductive glimpse of *eidos*—a form charged with wholeness and perfection—which still briefly precedes the act of exchange. It turns out that the negativity of objects is something we never forgive them for. The same can be said of those who work around the trash that all commodities necessarily become: in a world dominated by the secular perfectionism of fetish, the disgust they still arouse in middle class subjects is less an effect of uncleanliness than it is the intense revulsion they (we?) feel towards the negativity of the material itself.

In addition to all this, the *morph* of use again becomes the *hyle* of production: it is not simply that the object is the member of a class again; it is that it suddenly enters a state beyond determinacy. This is not a One become Many, but a Many become a qualitatively new kind of One. Twelve televisions on the back of a scavenger's cart are not simply the sum of twelve Ones, but a qualitatively separate entity, an invention. Objects displayed in such a way are less than objects: they are in some sense already the undifferentiated *material* they will return to (become) when processed by the recycler.

2

There is something inherently *critical* about the practice of the scavenging body. Critique, of course, in its classical sense substitutes essence for appearance, something real for something false. Though we have come to replace the residual Platonism of this formula with an understanding of critique as the construction of new problems (new sets of questions rather than definitive conclusions), there is a strictly performative or strategic dimension in which this opposition between essence and appearance continues to remain indispensable.

Critique, then, brings something hidden to light. Darkness, however, is unequally distributed. It exists in degrees, along continuums. One might say that every social order is fundamentally structured around an occluded hierarchy of secrets. Solving a murder is, from this angle, not critique but procedure: the secret it brings to light is mundane. This is not the case with the *shou polan de ren*. What they drag to light is not simply one object among others—a bare empiricist thing—but *an object which instantaneously discloses to the eye the logic of the whole itself*. Trash is precisely this microcosmic object, an object that in capitalist societies attains properly metaphysical dimensions, because it is imbued in some sense with *more reality* than others. Marcuse was one of the first to comprehensively map the feedback mechanism between late capitalism and garbage: he understood that garbage was not only an unintended consequence of capital, nor an externality or side-product, but its very *telos*.[3] In capitalist societies the production of new needs intersects with the disposability of commodities to form a socially necessary regime of trash. *Shou polan de ren* drag back into the light this absurdity at the heart of capital.

Of course, such a gesture cannot be confused with the quaint idea of "making something public." Yes, the object is withdrawn from the enclave of the domicile and placed back into the light of the *agora*; however, it is not returned to social consciousness by virtue of this passage.

Making something "public" could not possibly be an act of mere pointing. This is because the *res publica* is never just inertly shared space—this would perhaps better describe a market place—but a shared preoccupation, something created, an affair involving matters of ultimate concern to the community; which is to say that if the public is a space, it is not extensive but intensive. What the *shou polan de ren* drag through the city, even when they are close to its center, is always paraded on the margins. Three thousand plastic bottles driven through the city is an object lesson in capitalism's objective stupidity. There is nothing here to discuss or interpret, only something to do.

3

Critical attempts to foreground the ecological limits of productivism often take recourse to a strategy of scale. Edward Burtynsky's trash landscapes are here paradigmatic. Amanda Boetzkes has suggested that these works stage the erasure of nature without retreating into nostalgia for it: geographies of trash map uncomfortably onto the natural landscape they obscure and model, suggesting not only that nature is being screened out by production but that production itself is limited and defined by the unrepresentable nature it can't help but be. Boetzkes writes, "Nature is posited *through and against* garbage, as that which overflows the economic system but which touches the rubbish at its endpoint."[4] It may be that the politico-aesthetic activity of the scavenger operates as a supplement to Burtynsky's relentlessly aerial angle. There is, after all, a way in which the landscapes documented by Burtynsky threaten to naturalize productivism precisely by transforming catastrophe into necessity. The system of garbage becomes unlinked from historical materialist time and placed back into the intentionless *long durée* of stones: scale becomes the alibi for a politics reduced to looking—the austere, pseudo-neutrality of the document. There are no agents here, no histories, just the sublime interplay of forms visible from space. The aerial angle culminates in a picture of trash entirely subtracted from the cycle of shame mentioned above: a stark cleanliness of form, a pure monumental line, replaces the dark specificity of garbage—all of the rich vulgarity found in any square meter of human refuse.

Shou polan de ren encounter a mountain of garbage from below; it is not a topography but a smell. Seen from the right angle, they are masters of an art and science of bundles. In a society that equates time and money, every bundle is a gamble, a tentative, reversible pact established between mass and speed. It is logistics—a craft of form, space, and movement—that mediates the difference between a successfully completed run and the messy risk of the accident. We must remember that being noticed at all is a danger for someone living on the margins of legality: for some, a road full of spilled trash is much more than an inconvenience. It is in this domain of technique, experience, and risk that the virtuosity of the well-crafted bundle can be appraised. We should not be surprised if there is not an outright agonism at work here, an incidious art marked by wins and losses. A particularly bold construction becomes the immediate target of a competitor's doubt or praise—probabilities are calculated, stories told to memorialize the extremes. In a society less transfixed by status, the exquisitely conceived bundle would, no doubt, arouse the attention of adults, too. It is only during the experience of moving house that people today intuit the thoughtfulness of this art: the domicile is suddenly encountered not as effortlessly traversed space, but frustratingly packed cubic volume: one's objects are no longer invisibly used, light as

Andrew Pendakis, *Shanghai, China*, 2013.
Color photograph.

air, but incredibly dense and heavy. Objects experienced in this way are at once irreducibly specific (*this* edge, *that* corner) and oddly generalized—indeterminate stuff whose primary quality is pure weight. Mass functions here as the antidote to the austere scientism of Burtynsky's privileged

Andrew Pendakis, *Shanghai, China*, 2013.
Color photograph.

aerial gaze: the limit it places on will is the point of departure for an historical ontology of labor, a Hegelianism grounded in (collective) muscle. However small the dent it makes in the mountain of waste, every bundle produced by the scavenger is a cipher for the human capacity to differentiate itself from stone.

We know after Foucault that space, like politics, is war by other means. We should not hesitate, then, to note the ways in which the cart of the scavenger doubles as an instrument of war. For an instant, the everyday indignities of political smallness are traded in for the potency of sheer mass. Understood from the context of a society in which some drive cars the price of houses and others lack a place to sleep how could there not be a middle finger or two hidden somewhere in these bundles? In the absence of genuine recognition sometimes being seen at all is a disguised kind of authority. Which is only to say that the Lexus that pauses in the shadow of a scavenger's precarious pile catches, if only for a moment, an objective glance of its own place in the system of production. Perhaps, then, the well-crafted bundle is not the strongest, not the one whose bindings best fasten it to itself, but the one that comes as close as possible to falling without actually doing so.

4

The scrap collector entangles time, cross-listing eras in a way that can be mystifying. How, for example, would one locate such a figure on the standard dialectical calendar: is she residual, emergent, or eternal? On the one hand, we are clearly in the presence of a past. If her cart has a motor, her hands and strategies remain analog (which is not to say uncreative). This is an action as old as gleaning: stooping over piles of motherboards is surely a distant cousin of crouching over wild mushrooms. Scavenging is as old as the difference between the animal and the mineral. Indeed, the habits of the scavenger make direct contact with the history of mammals; these pretentions to progress are placed back into the appetites of the hungry body we share with birds and lions. Of course, it is this proximity to the universal opportunism of eating that makes the figure of the scavenger available to ideological expropriation by survivalisms of every kind and to logic that reduces existence to Hobbesian competition. If the classical proletarian body imagined by socialist realism was active, muscular, lit up by the red of a blast furnace, the *shou polan de ren*, like the *lumpenproletariat*, are bodies useful to the myth of transcendental human frailty—bodies used to mythologize weakness as constitutive of things themselves. Christianity finds solace in this figure of poverty that without being transformable at least remains "resourceful": this is precisely the horizon on which it stages its many humilities and pities, the *mise-en-scene* of any working *caritas*.

But is it also not the case that the scrap collector, perhaps more than we can imagine, belongs less to the past than to the future? From *Mad Max* to *The Road*, *The Walking Dead* to *Wall-E*, our culture seems to broadly intuit the future not as a space of techno-convenience and happiness, but as ruins, mass scarcity, and violence. The future envisioned by these texts is one in which scavenging is the only form of productive labor left: it no longer takes place in the interstices of capital but openly replaces it. When Franco "Bifo" Berardi says that the future will be characterized by "frugality" he invokes not a horizon of catastrophe but one of self-imposed limitation, an abundance made out of wanting less. Frugality can arrive in the form of a disaster, or it can exist as the target of a politics; it can befall us, or it can form the affective basis of a plan.[5]

To see the present from the perspective of the scavenger is to see every whole as a tentative ecosystem of moveable parts and pieces. Everything becomes decomposable, alterable—charged with alternate usages. *Shou polan de ren* never participate in what Nietzsche once called "the melancholy of everything finished."[6] Only machines and houses seen from one angle—that of capital—depreciate: everything else fructifies.

It needn't even be said that those who sentimentalize this craftwork of the poor are as guilty as those who ignore it. Walter Benjamin knew, after all, better than anyone that there is no tension between celebrating what exists and envisioning a world in which it doesn't.

Notes

1 Aristotle, *Physics* (Lincoln: University of Nebraska Press, 1961), 91.

2 Richard Shears, "An engagement ring-ring! Chinese Romeo spends £55,000 on 99 iPhone 6s to impress a girl he wanted to marry … but she says No," *Daily Mail*, November 11, 2014. http://www.dailymail.co.uk/news/article-2829815/An-engagement-ring-ring-Chinese-Romeo-spends-55-000-99-iPhone-6s-impress-girl-wanted-marry-says-No.html

3 Herbert Marcus, *One Dimensional Man* (Boston: Beacon Press, 2012), 49.

4 Amanda Boetzkes, "Waste and the Sublime Landscape," in RACAR *(Revue d'art canadien/Canadian Art Review)* 35.1 (2010): 30.

5 See Andrew Pendakis, "Franco 'Bifo' Beradi and the future of capitalism: 'We have to run along the line of catastrophe'," in Todd Dufresne, ed., *The Economy as Cultural System: Theory, Capitalism, Crisis* (London: Bloomsbury Press, 2012), 169-75.

6 Friedrich Nietzsche, *Beyond Good and Evil* (Cambridge: Cambridge University Press, 2002), 169.

3 *The Ethics of Earth Art* as Catalytic Theory Converter

Mark A. Cheetham

I do most of my reading on a computer screen these days. Even paperbacks seem limiting since they don't as readily afford me the luxury of comparing texts simultaneously on my laptop. I make an exception for my copy of *The Ethics of Earth Art*, not only because it came from the author, but because it's dog eared with marginalia and sticky notes (the latter a failed remedy for not defacing my books). I read the book when it came out in 2010. I've used it as a text in graduate courses. I keep recommending it because I think it is the strongest account of earth art in a healthy and growing field of publications on this subject. Why? I admire how well Amanda Boetzkes uses philosophical theory in her articulations of artworks. I don't mean that she *applies* Martin Heidegger's, Luce Irigaray's, or John Sallis's thinking accurately and productively to works of earth art, to their paradoxical revelations of the withdrawal of the earth from anthropocentric comprehension for example. An instrumental use of theory of this sort is often the default in the field of Art History, which too often assumes that "the work" comes first, innocent of theory. My sense is that instead, Amanda Boetzkes construes theory—which for me is a more capacious designation than Philosophy or Aesthetics, much as Visual Culture and its cognates is a more inclusive field than Art History—as integral to artworks and to their interpretation. Whether she is analyzing Amelia

Jones's work with Merleau-Ponty's phenomenology in the context of Ana Mendieta's *Silueta*, or Craig Owens' deployment of Walter Benjamin's theory of allegory to understand Robert Smithson's earth works, Boetzkes does not pretend that one supposedly self-contained realm is seen in terms of the other, whether in a relation of influence or as a key to understanding. Her encapsulation of the import of Smithson's *Spiral Jetty* (1970)—deservedly the most renowned example of earth art, it exists in three iterations: sculpture, essay, and film—can stand for the relation of theory to artwork throughout *The Ethics of Earth Art*: Smithson's "conceptualization of *Spiral Jetty* braids the site and the text together and circles them around points of disruption that correlate to both an expulsion of dynamic substance the loss of discursive coherence."[1] Building on the ideas of Deleuze and Guattari, Elizabeth Grosz writes in this regard that "philosophy may have a place not so much in assessing art (as aesthetics has attempted to do) but in addressing the same provocations or incitements to creation as art faces."[2] For Boetzkes, that fundamental call comes from the elemental status of the earth. Neither artworks (as image and material) nor texts (philosophy and theory more generally) are origins; all are fragmentary, partial, and to use a popular term these days, entangled. Boetzkes is unusual, though not unique, in interweaving rather than separating these components. By coincidence, when asked to contribute to this publication I was reading Christine Ross's truly profound book *The Past is the Present; It's the Future Too: The Temporal Turn in Contemporary Art* (2012). Months later, I'm still reading it; the material is complex and compelling in its understanding of theories of time in relation to recent art. Ross was Boetzkes's doctoral supervisor at McGill

University. Both scholars read theoretical issues and texts *with* contemporary artworks in ways that suggest that these works are themselves theoretical. For me, this is the best way for art historians to examine adequately such difficult and thematically important works of art and perhaps also to explicate the theoretical texts.

The invitation to contribute to *Plastic Blue Marble* has taken me back yet again to *The Ethics of Earth Art.* The book continues to inform and challenge me as I work on my parallel project, *Landscape into Eco Art: Articulations of Nature since the '60s.* There is a tendency in recent literature and exhibitions that examine what I call contemporary eco art—Boetzkes instead extends the term earth art from the 1960s context to the present—to describe individual works' presence, effects, and evolution without saying much about what we might call the geoaesthetic theory that informs many of these works. Again, the theory-rich discussions that are at the core of Boetzkes's book are rare.[3] Vitalized by her departure from the norm, I want to outline here two theoretical tacks that aided me in understanding eco art. The first adapts Theodore Adorno's much cited judgment that it is barbaric to produce poetry after Auschwitz. The second theorizes pioneer eco artist Hans Haacke's comment—quoted in *The Ethics of Earth Art*—that his environments from the 1960s and 1970s sought to "articulate something natural" for the artist and his audience.[4]

Rapid climate disruption and its increasingly serious consequences worldwide encourage many of us concerned with contemporary art to ask an old question with renewed urgency: what can and should artists and visual art do in the face of these pressing planetary problems? While not necessarily posing such a grand question, recent eco art practices provide a range of possibilities, as does Boetzkes's book. Some are ameliorative in themselves, such as land reclamation projects. Mel Chin's *Revival Field*, 1991-93, is the best-known example; on a much larger scale is Viet Ngo's *Devil's Lake Waste Water Treatment Plant in North Dakota*, 1990. Others are calls to awareness, such as the alarming photographs of environmental degradation by David Maisel and in Maya Lin's *What is Missing*, a melancholy interactive website begun in 2009 that documents what many are calling the sixth mass extinction of life on planet earth. These and related artistic responses are as individual as they are planetary in implication, as numerous as the concern with climate is enormous, and often as material as they are embroiled in both cultural and scientific ideas. A question that arises around some eco art is whether the works should be considered art as opposed to technology or social activism, and whether that designation matters.

We can adapt two of Adorno's arguments to gain a general perspective on visual art's relationships with climate change. In *Aesthetic Theory* (1970), Adorno held that if art is to remain connected to momentous societal problems, it must fight for an identity distinguishable (if not fully autonomous) from its culture. On this view, Chin's *Revival Field* and the German artist Herman Prigann's (1942-2008) many land restorations in Europe must function as art as well as leach toxins from the soil or repair unsightly and toxic strip mines. As Adorno wrote, "All efforts to restore art by giving it a social function … are doomed."[5] In his terms, "Art's double character as both autonomous and *fait social* is incessantly reproduced on the level of its autonomy."[6] Art must be identifiable as such if it is to have an effect.

There are many possible objections to this stance. Malcolm Miles claims that eco art "crosses boundaries between art, social research, and environmentalism so that it no longer matters whether it is art or something else."[7] He suggests, hopefully, that "If the aim is to shift the balance of humanity's relation to the earth from exploitation to sustenance, this implies a shift in human relations as a point of departure … An ecological aesthetic [can be seen as an] intervention in social conditions, seeing human nature not in a biological sense as beyond history, but as produced in history … Art can intervene in writing the scripts, interrupting the processes of normalization."[8] His prime example is Mierle Laderman Ukeles's *Touch Sanitation*, 1974-84, a now legendary project in which the artist drew attention to the urban sanitation systems that we unheedingly depend upon by shaking the hand of every sanitation worker in NYC. But his point, while in keeping with the fundamental ecological premise of the interconnectedness of all phenomena, is precipitous. Granted, what humans call a given eco-aesthetic project doesn't matter to what we call nature. But our naming and categorizing practices do matter profoundly to us: they influence, if not determine, what we see and how we act, as the histories of the over-determined concepts of "art" and "nature" attest. An example relevant to my discussion of Hans Haacke below was related by author Jack Burnham in 1967:

> I can remember when Haacke took me to see an example of his first water boxes (spring 1962), then in the rental collection of the Museum of Modern Art in New York. A secretary commented that the museum personnel had been playing with

> it for days—it seemed to have caused more joyful curiosity than any number of "sculptures"—for that reason the museum never thought seriously of buying it as a "work of art."[9]

Adorno wrote powerfully on nature and natural history, but my intention is not to engage with his thinking per se but rather to pose a version of his famous challenge to the aesthetic as it operates in the world. My argument adapts Brian A. Oard's gloss of Adorno:

> To persist, after Auschwitz, in the production of monuments of the very culture that produced Auschwitz (Adorno might have spoken of Strauss's *Four Last Songs* rather than generalized "poetry") is to participate by denial in the perpetuation of that barbaric culture and to participate in the process (reification) that renders fundamental criticism of that culture literally unthinkable.[10]

I can specify my emphasis on the "art" in eco art by asking if it is legitimate to recast Adorno's famous question about the authenticity of artistic expression after humanity's nadir at Auschwitz in light of contemporary ecocide. Can we justifiably make art about nature in full cognizance of anthropogenic climate disruption? Can eco art persist in its creation of objects and interventions in the face of humanity's undeniable causation of global climate change—what commentators are variously calling the Anthropocene (Crutzen), the Capitalocene (Haraway), or Anthrobscene (Jussi Parikka's term, calculated to underline the obscenity of the wanton disregard for and humiliation of integrity, whether the earth's, of humans, of non-human animals and other living and also inanimate materials).[11] Should we continue to produce works and to display them using the same largely capitalist structures and even attitudes that spawned our current climate problems? I would say yes, if eco art's effects lie in reflecting and modifying the longstanding relationships between artistic expression, landscape, and human views of the earth and nature. With Adorno again, it is only from the sometimes nominal remove that defines it as art that eco art can meaningfully speak to our current climate predicament.

Amanda Boetzkes is less concerned with the status of instrumental art reclamation than with what these initiatives (and the earth art that she discusses) reveal about the earth's ultimate unavailability to human perception, how the earth exceeds what we can perceive and retracts itself from us in its ineluctable difference. This challenging position is fundamental to her book. "The artwork is the threshold at which elementals exceed the limits of perception," she claims. "In simultaneously making contact with natural phenomena and withholding the drive to unify them in the viewer's field of vision, the artwork offers itself as a medium on which the earth manifests and asserts its irreducibility to human signification."[12] Thus she reads recuperative projects by Betty Beaumont, Rebecca Belmore, Joseph Beuys, Basia Irland, and Aviva Rahmani against the artists' routine statements of ameliorative purpose, accentuating instead the works' recessive ethics, defined as "a stance of retraction from and receptivity to the earth that foregoes the propensity to actively subsume it within the parameters of our existing logic."[13] In her persuasive argument, "artists create the conditions of possibility for the earth to appear at the limits of intelligible form …".[14]

In developing pioneer ecological artist Hans Haacke's drive to "articulate something natural" in his systems-based eco art of the 1960s and 1970s, I see eco art as an aesthetic "articulation" of evolving ecological issues. If one way to avoid eco art's absorption into the industrial society that spawned the Anthrobscene—to insist on its status as art rather than something instrumental—is to understand its potential relationships to this phase of the earth's life in earthly and technological systems and in the human and the non-human spheres, then Haacke's installations are a worthy test case. They were among the first that we could call ecological in the sense that they were functioning illustrations of the interrelatedness of systems in nature and of our own technological relays, for example in the case of *Rhine Water Purification Plant*, water filtration and fish life. Installed in the Museum Haus Lange in Krefeld, Germany, in 1972—and the cause of an uproar when it was revealed that a source of the serious pollution of the river water that the work filtered was the sewage treatment facility funded by the same civic authority that supported the gallery—Haacke's demonstration was clearly serviceable. The fish seemed to thrive. But he clearly established too that his display was not a version of the inadequate Krefeld waste plant: he drew the still-polluted Rhine water downstream from that facility, after it had been treated by the "real" industrial process. He also placed his work in an art gallery, a frame for aesthetic reckoning. Haacke's stated goal was to "articulate something natural."[15] For Haacke as for Adorno, we could argue, "the artwork is related to the world by the principle that contrasts it with the world."[16] To function as an articulation of what Bruno Latour would call "concerns" that bring human and non-human actants, plus scientific, technological and aesthetic discourses into contact, an artwork must remain distinct from the cultural contexts whose materials and habits it unavoidably employs.

While a fine line can be seen to exist between art and non-art activities in general and between eco art and green technology projects, for example, it is a malleable boundary, an edge in flux for any number of reasons. Eco art can articulate this border, not to guard it, but rather to monitor what its inevitable shifts imply for humans, other animals, and perhaps in Boetzkes's terms, the earth as it recedes from our view. We can probe these speculations through her reading of Haacke's 1965 *Condensation Box*. In this and related works, "Haacke activated natural processes within the gallery space,"[17] in this case, the weather-like effect of condensation. As Caroline A. Jones reminds us in "Hans Haacke 1967"—her masterly article accompanying the recreation of Haacke's 1967 exhibition at MIT—the later version of *Condensation Cube* exhibited there was titled *Weather Cube*.[18] Boetzkes claims that "Haacke's work sensitizes the viewer to the otherwise hidden dimensions of natural activity .... His practice invited the unpredictability and fundamental impenetrability of the elemental".[19] While I am convinced by this claim, especially given the consistency with which Boetzkes demonstrates the thesis across many otherwise different works of earth and eco art throughout her book, it is for me not the whole story. For example, it does not take account of Haacke's attempt in this and cognate work to create autonomous systems that are as much as possible independent of human input and even observation. Claiming that "rather than presenting the spectator with information (scientific facts about ecosystems or environmental degradation)," Boetzkes writes that Haacke's ecological work instead "requires the spectator to stand, watch, and wait for elementals to reveal themselves."[20] While viewer participation is finally important to these installations, Jones applies more sustained pressure to Haacke's will to achieve what I think of as an Adorno-like autonomy. To Burnham, the artist wrote that "in spite of all my environmental … thinking I am still fascinated by the nearly magic, self-contained quality of objects. My water levels, waves, and condensation boxes are unthinkable without this physical separation from their surroundings."[21] Jones emphasizes the ironic tension in "how Haacke struggled to keep the human from impeding the autonomy of these fluid systems, yet recognized the importance of the art in restoring humans' own equilibrium (via empathetic 'systems' he was not acknowledging as part of his concern)." She concludes, "The human could watch; the human might even push a system into motion, but the system's unfolding was independent of the human in 1967."[22] With Boetzkes, we might conclude that what is definitively independent of the human is the earth, here seen in its difference through eco art. But such a recession is not all we can potentially realize when art claims to "articulate something natural."

I am not suggesting that Haacke meant more by "articulate" here than two of the *OED*'s standard definitions indicate: "To express distinctly" and "to attach or unite." Given that "articulation [was] perhaps one of the most generative concepts in contemporary cultural studies"[23] in the 1980s and 1990s, however, we might well extend its theorization at the hands of Ernesto Laclau and Stuart Hall particularly to think with it in the context of eco art. In Slack's exegesis, for Laclau, while articulations do link concepts, such relationships are never necessary and do not connect systematically across a system.[24] Laclau, and Hall after him, work against any deterministic system by insisting on the specificities of articulation. Hall posits that the notion of articulation "has the considerable advantage of enabling us to think of how specific practices articulated around contradictions which do not all arise in the same way, at the same point, in the same moment, can nevertheless be thought together."[25] If I may lift this idea shamelessly out of the contexts in which Hall himself would deploy it—though not away from the timeframe he shared with Haacke's work—I can build on Hall's admirably clear and evocative idea. "In England," he stated in an interview:

> the term has a nice double meaning because "articulate" means to utter, to speak forth, to be articulate. It carries that sense of language-ing, of expressing, etc. But we also speak of an "articulated" lorry (truck): a lorry where the back (trailer) can, but need not necessarily, be connected to one another. The two parts are connected to each other, but through a specific linkage, that can be broken. An articulation is thus the form of the connection that can make a unity of two different elements, under certain conditions. It is a linkage which is not necessary, determined, absolute and essential for all time. You have to ask, under what circumstances can a connection be forged or made?[26]

The linkage that Boetzkes posits between Haacke's *Condensation Cube* and a showing forth of the earth is highly plausible, but neither she (I imagine) nor I would say that this is the only articulation performed by this example of eco art.

My own thinking and writing on eco art has been impelled and challenged by Amanda Boetzkes's book in many ways. Of the stimuli I'm conscious of, her concluding section, "Facing the Earth Ethically"—which focuses on art that uses the "elemental" water—has become central to my own research on affect in eco art. Her comment that the generation of originary, 1960s earth artists and those practicing in related ways today "are connected by their elucidation of the earth as a domain of ethical concern"[27] —I realize well after the fact—got me thinking about the intersections and divergences between generations of artists concerned with what I call eco art, especially as this entire five decade history is encapsulated in the recent work of artists who were there at the start and have continued to make earth art, such as Michael Heizer or the late Nancy Holt. A question that arises for me when thinking about the catalytic effect of this book is how much one also writes against work that is so clearly influential to one's own ideas. While I usually agree with Amanda Boetzkes's analyses, I do also sense my resistance to her oft-voiced claim that land, water, and other elementals "reveal the earth's resistance to form, image, and structure."[28] I concur that this is often the case but would suggest that there are also important exceptions and other narratives. For example, unlike Boetzkes and a host of other commentators recently, I want to make room in the discussion of eco art for earlier Western "landscape." This genre should not be dismissed as quickly as it is here, and as with contemporary eco art practices, its focus was not always on elementals.

"Earth art resists delivering nature as a thematic image, such as a landscape, or a tangible object, such as a specimen in a natural history museum," Boetzkes claims.[29] I would counter that even some traditional Western landscape paintings are not thematic in this restrictive sense. Perhaps she is thinking of the endless views of this or that place or of hackneyed picturesque effects. No doubt the tiresomeness and limitations of this type of landscape spurred the '60s generation to do something quite different. Michael Heizer, for instance, asserted that looking directly at the land was "more interesting than looking at works in the Louvre or Metropolitan."[30] Cultural theorists have been equally dismissive, notably W. J. T. Mitchell in his influential collection *Landscape and Power*, who declared in 1994 that "Landscape is an exhausted medium, no longer viable as a mode of artistic expression."[31] If that's now old news, the prominent theorist Timothy Morton also needs to diminish landscape in his recent description of his category of *hyperobjects*, entities "so vast, so long lasting, that they defy human time and spatial scales," adding that "They wouldn't fit in a landscape painting."[32] Of course conventional landscapes are limited; so are all human art forms, according to Boetzkes. Can't framed landscapes also intimate this very limit, as the 18th-century discourse of the sublime does explicitly? The point of my question is that we should not foreclose on the landscape tradition so quickly.

The advent of "Eco art history"—defined in a College Art Association of America session in 2014 as an initiative designed to "bring together art historians from diverse fields to work toward a more earth-conscious mode of analysis"—suggests to me that those of us concerned with ecological concerns in contemporary art could also beneficially be looking to earlier art's articulations of the earth. Let me conclude in this spirit with landscape that would have no place in *The Ethics of Earth Art*, J.M.W. Turner's renowned

Snow Storm—Steam-Boat off a Harbour's Mouth…,[33] exhibited in 1842. To the bafflement and outright hostility of most of its early viewers, this dramatic work places us in the roiling sea, from which we precariously view a struggling steam boat firing a distress flare as it fights to make safe harbour during a storm. Whether or not Turner was lashed to the mast of such a boat to experience and record the elements, as he claimed, is ultimately immaterial. He fully conveys the chaos of the storm: the horizon is dangerously off level and we cannot distinguish water from sky from spray. Turner's famed indistinctness of form ensures that there is much we cannot see here. Our senses are overwhelmed and inadequate, suggesting his interest in the sublime and offering an opening for an argument about elementals exceeding our grasp and intimating the presence of the earth's forces. We can be historically specific on this point. James Hamilton claims that Turner was conveying in the design of the painting the powerful forces of magnetism harboured by the earth, those discovered by Michael Faraday and widely disseminated by the scientist—and Turner's close friend—Mary Somerville.[34] We can see the effects of this energy in the vortex but we cannot see this elemental force itself. My reading would not, however, end with a recognition of elementals—not impossible in Turner—and focus instead on the longer history of the earth's articulation. While this picture is framed, everything about it breaks free of the landscape genre's conventions. A person of his time rather than ours – which is to say of the Industrial Revolution – in this and many other works, Turner celebrated and sought to convey the affective impact of our technological innovations while simultaneously witnessing the power of nature. For him these forces co-exist and are too closely intertwined to be conveniently opposed in a narrative of struggle. Turner would not likely have agreed with his ardent promoter John Ruskin that "The interest of a landscape consists wholly in its relation either to figures present—or to figures past—or to human powers conceived. The most splendid drawing of the chain of the Alps, irrespective of their relation to humanity, is no more a true landscape than a painting of this bit of stone."[35] Yet because we are implicated as participants—not mere observers—in this drama and in Turner's other historical landscapes, the earth or nature is not presented as an independent force, but always with its human interlocutors. It is only with the post-human interests of some contemporary eco art that this default emphasis is challenged.

Notes

1 Amanda Boetzkes, *The Ethics of Earth Art* (Minneapolis: University of Minnesota Press, 2010), 83.

2 Elizabeth Grosz, *Chaos, Territory, Art: Deleuze and the Framing of the Earth* (New York: Columbia University Press, 2008), 2.

3 Malcolm Miles's *Eco-Aesthetics: Art, Literature and Architecture in a Period of Climate Change* (London: Bloomsbury Academic, 2014) is another welcome exception to the pattern of description, especially in his extended use of Herbert Marcuse's thinking in relation to ecological art and its societal issues.

4 Boetzkes, *The Ethics of Earth Art*, 45.

5 Theodor W. Adorno, *Aesthetic Theory* (1970), eds. Gretel Adorno and Rudolf Tiedemann, trans. Robert Hullot-Kentor, (Minneapolis: University of Minnesota Press, 1997), 1.

6 Ibid., 5.
7 Malcolm Miles, "Aesthetics and Engagement: Interested interventions," in *Ecological Aesthetics: Art in Environmental Design: Theory and Practice*, ed. Heike Strelow (Basel: Birkhäuser, 2004), 202.
8 Ibid., 206.
9 Jack Burnham, "Hans Haacke, Wind and Water Sculpture." in *Art in the Land*, ed. Alan Sonfist (New York: Dutton, 1983), 109-110.
10 Brian A. Oard, "Poetry after Auschwitz: What Adorno Really Said, and Where He Said It," Mindful Pleasures (blog), accessed December 18. 2015 http://mindfulpleasures.blogspot.ca/2011/03/poetry-after-auschwitz-what-adorno.html.
11 Jussi Parikka, *The Anthrobscene* (Minneapolis: University of Minnesota Press, 2015)
12 Boetzkes, *The Ethics of Earth Art*, 21.
13 Ibid., 4.
14 Ibid.
15 Hans Haacke, quoted in Jack Burnham, and Hans Haacke, *Hans Haacke, Wind and Water Sculpture* (Evanston, Ill: Northwestern University Press,1967), 113.
16 Adorno, *Aesthetic Theory*, 7.
17 Boetzkes, *The Ethics of Earth Art*, 44.
18 Caroline A. Jones, "Hans Haacke 1967," in *Hans Haacke 1967* (Cambridge: MIT List Visual Arts Center, 2011), 6-26.
19 Boetzkes, *The Ethics of Earth Art*, 45.
20 Ibid., 46.
21 Burnham, *Wind and Water*, 120.
22 Jones, *Hans Haacke 1967,* 14.
23 Jennifer Daryl Slack, "The Theory and Method of Articulation in Cultural Studies,"in *Stuart Hall: Critical Dialogues in Cultural Studies,* eds. David Morley and Kuan-Hsing Chen (New York: Routledge, 1996), 113.
24 Ibid, 120.
25 Stuart Hall, quoted in Slack, "The Theory and Method of Articulation in Cultural Studies," 123.

26 Lawrence Grossberg, ed."On Postmodernism and Articulation: An Interview with Stuart Hall." *Stuart Hall: Critical Dialogues in Cultural Studies*, 141.

27 Boetzkes, *The Ethics of Earth Art*, 3.

28 Ibid., 16.

29 Ibid., 12.

30 Michael Heizer, quoted in Germano Celant, *Michael Heizer* (New York: Art Pub Incorporated, 1997) 62.

31 W. J. T. Mitchell, ed., "Imperial Landscape" in *Landscape and Power* (Chicago: University of Chicago Press, 1994), 5.

32 Timothy Morton, "Zero Landscapes in the Time of Hyperobjects," *Graz Architectural Magazine*,July, 2001, 80.

33 The painting's full title: *Snow Storm - Steam-Boat off a Harbour's Mouth Making Signals in Shallow Water, and going by the Lead. The Author was in this Storm on the Night the "Ariel" left Harwich*

34 James Hamilton, *Turner* (New York: Random House, 1997), 355.

35 John Ruskin, "Lectures on Landscape," 1870, accessed December 27, 2015. http://www.readbookonline.net/read/19610/56151/

4 Astro Was Here

Maria Whiteman

1. Perception and Interpretation

The close reassessment of how one perceives does not take place in a vacuous space; it occurs through elementals.[1]

Sensation is enabled by the distinctly recessive orientation that the space mediates.[2]

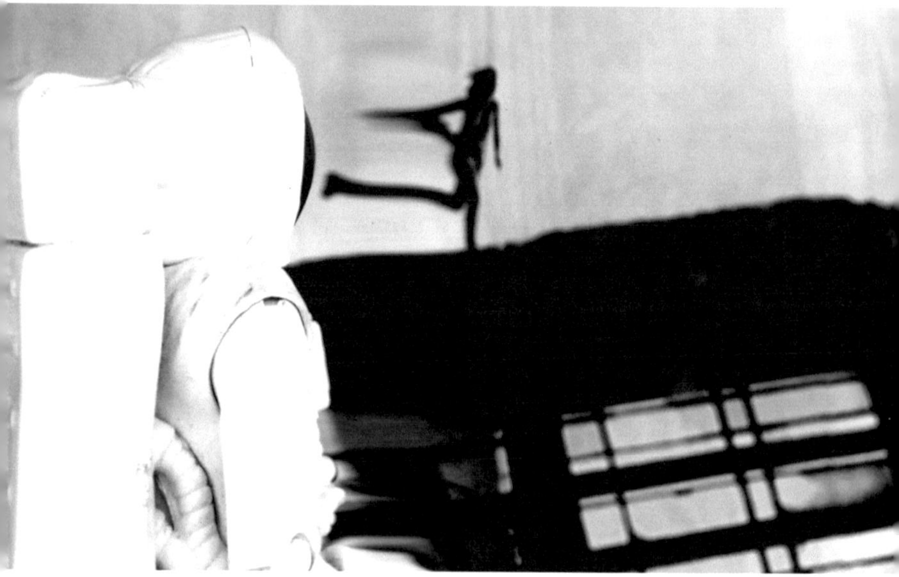

The human world is delayed; the environment is not divided into graspable things and instead is reconceived in terms of elementals that close against the senses. It is this closedness, or "disconcealment," of the environment that makes the sky appear through the aperture as opaque.[3]

Interpretation—not simply perception, but the response and the possibilities of response—occurs within the optic array, through the allure of objects, the vicarious transmissions between persons, real objects and surface effects.[4]

The interpretations of objects are non-sequiturs; gestures and signals that bring both the body and the object into visibility out of the concealment of equipmentality through their co-extensive nonfunction. They disclose no use or meaning.[5]

2. Ecotechnology

The project is based not on an anthropocentric system of resource management but rather a fundamental shift of consciousness, both ethical and aesthetic. This shift entails a retraction, not simply a literal withdrawal from the land but a fundamental recession that stems from the acknowledgement that human necessity is preceded by, but inextricable from, the earth's imperative.[6]

Technology, which is always a supplement and therefore a departure from the presumed stable and pure condition of nature, actually precedes what we understand nature to essentially be.[7]

3. Wasting Nature

> Nature as such cannot be seen but nevertheless behaves as an inextricable and vital force that impresses itself into the screen of waste by stretching and tearing at it. Nature deforms the barrier of garbage that sublimates it out of sensibility. ... in its undoing of garbage and the visual limit that the screen of waste demarcates, nature registers its own disappearance from sense.[8]

The concealment of nature in trash and the closure of vision are co-extensive, thus disrupting the analogy between wild nature and the freedom of the mind.[9]

If the hyperproduction of garbage is a symptom of the spread of ecotechnology that reinforces human dominion over the planet, then the sublime landscape of waste enacts the loosening of this suffocating agglomeration. It is precisely the procedure of undoing or unworlding that is necessary to reinvent a relationship to the earth as such. What is at stake, then, is not merely a new visual language of nature, but a release from the constrictions of anthropocentric discourse altogether.[10]

4. Earth and Elementals

The sky, in its seemingly unlimited height and expanse, formulates a horizon in its overlap with the earth...[11]

In this intermediary zone between elementals, light becomes almost palpable as it pours into the chambers…[12]

The elemental is crucial here, for there is no facing that is not understood through sensible qualities. The elemental yields the face, and itself surfaces in the gesture of facing.[13]

Natural processes become a primary language to which we are privy before human language.[14]

What deserves consideration are the questions of what it means to disclose the earth as elemental and how this demands a specifically withdrawn and receptive mode of confrontation.[15]

The human touch is understood as ornamentation, a word that presumes that this contact is a superfluous addition to the earth, and hence a technique. Through habit and ritual, the superadded action of forming base material produces and naturalizes the earth itself.[16]

5. The Ethics of Encounter

Is it possible to advance ethical imperatives in and through the belatedness of ecology?[17]

The other, and we might say the otherness of the earth, are closed off in the tight weave of an all-encompassing grip.[18]

Image Credits

Maria Whiteman, from the series: *Astro Was Here*, 2015-16. Photographs.

Notes

1. Amanda Boetzkes. *The Ethics of Earth Art*. (Minneapolis: University of Minnesota Press, 2011), 119.
2. Ibid.
3. Ibid, 122.
4. Amanda Boetzkes. *Heidegger and the Work of Art History*. (Burlington: Ashgate Press, 2014), 287.
5. Ibid.
6. Boetzkes. *The Ethics of Earth Art*, 200.
7. Amanda Boetzkes. "Waste and the Sublime Landscape," *RACAR (Revue d'art canadien/Canadian Art Review)* 35.1 (2010), 30.
8. Boetzkes. "Waste and the Sublime Landscape," 31.
9. Ibid, 28.
10. Ibid, 31.
11. Boetzkes. *The Ethics of Earth Art*, 106.
12. Ibid, 113.
13. Ibid, 161.
14. Ibid, 172.
15. Ibid, 102.
16. Amanda Boetzkes. "Techniques of Survival: The Harrisons and Environmental Counterculture," *West of Center: Art and the Counterculture Experiment in America, 1965-77*. Eds. Elissa Auther and Adam Lerner. (Minneapolis: University of Minnesota Press, 2011), 311.
17. Ibid, 319.
18. Ibid, 159.

5 Ecologies of (Imaginary) Friendship

Ted Hiebert

In August 2015, the media world went wild over a report that the sound of drones makes black bears anxious—reported by scientists who measured the heart rate of a group of bears while flying unmanned aerial vehicles over their heads. The irony of the study is that in order to measure the effects of the drones, it was necessary to technologically upgrade the bears by outfitting them with cardiac monitors and bio-trackers designed to measure, record, and share the real time physiological responses of the animals to their environment.

Around the same time as this story broke, I was reading Amanda Boetzkes's *The Ethics of Earth Art* along with some of her other works. The story of the black bears does not at first seem like a work of art, but it struck me that there was something resonant between Boetzkes's process of inquiry and the scientific experiment. There was something missing from the analysis of the bears that Boetzkes's reflexivity helped me understand, namely the way the experiment implicated itself (or failed to implicate itself) in the causal matrix it was designed to study. The conclusion of the scientists was that drones make black bears nervous (and therefore flying them around wild animals should be more heavily regulated), but I couldn't shake the feeling that the study was designed with nervousness in mind and with a decidedly human sense of entitlement to operate on the natural world. Something in the experiment refused to be explained by the scientists' data however, something perhaps "unrepresentable," to use Boetzkes's term for a type of concept that stands in for the inability to represent what lies beyond the limits of representation. For the differing registers of scientific intervention seemed to uncomfortably ignore the actual experience of the bears in ways that instrumentalize—and thus also risk trivializing—the ecological conscience that the study advocates.[1] To put it differently, something in the story of the bears seemed to persist in excess of the data, as irreducible to the data as it was ambivalent to the good intentions of the scientists, something that perhaps cannot really be measured, but can certainly be (at least partially) imagined. In other words, something aesthetic.

This essay meditates on the story of digitally enhanced black bears for the ways that it might be understood differently, using Amanda Boetzkes's *The Ethics of Earth Art* as a catalyst for speculating about the ways that contemporary (anthropogenic) life problematizes the boundaries between science, ecology, and art. Indeed, if the story of these digital black bears is any indication, one might suggest that the evolution of technological culture is increasingly just as amenable to aesthetic analysis as it is to the scientific distillation of data. In fact, one might go as far as to suggest that the tools of art criticism have never been more relevant than at a time of widespread human impact on the environment: when it becomes eerily unclear how to examine and engage with the aesthetic implications of scientific discourse or of human culture as a whole.

Making Black Bears Anxious

The study in question was initiated by Mark Ditmer, a wildlife ecologist from the University of Minnesota St. Paul. In the experiment, Ditmer and his team tagged, tracked, and monitored four black bears living in a state park in Minnesota.[2] To gather the data needed, the bears were outfitted with GPS collars (biologger collars) and surgically implanted with heart sensors—human-grade cardiac monitors that were customized to the physiology of the bears. Once processed, the bears were released back into the wild to resume their natural patterns of behavior. Shortly thereafter the scientists began to fly small drones overhead, purposefully maintaining a distance from the bears in order to register more as background noise than as immediate threat. Meanwhile, Ditmer and his team recorded the heart rate variations that the presence of the drones elicited in the animals. The bears showed no overt behavioral signs of distress, but there was a dramatic acceleration of their heart rates, which in most cases doubled—and in the case of one of the bears, nearly quadrupled. Ditmer summarizes:

> In all 17 drone encounters for which the researchers had heart-monitor data, the bears' heart rates soared—one by as much as 46 beats per minute, another by 56 beats per minute, and a third (a mother bear with two cubs) by more than 120 beats per minute, from 41 to 162. While their heart rates returned to a more normal resting rate within 10 minutes after the encounter, the drone's presence clearly caused the animals at least some significant physiological stress.[3]

To begin the task of looking aesthetically at the study would be to note that there is something undoubtedly eerie about the idea of these cybernetically enhanced black bears—bears with technological prostheses that allow a strange form of communication, perverse though it may be, with the scientists on the other end of the data exchange. The bears are made to speak in the language of data, through their surgically implanted communication upgrades. The drones become the voices of the scientists in return, creating a biofeedback loop through which people make their presence known by casting technological shadows on the animal world and the bears speak back through a language of physiological distress. That these are cyborg animals is not really anything new—this is a relatively predictable manifestation of a scientific imagination that cares more about registering data than it does about thinking through the paradoxes of its processes. The details are in the particularities however, and one need not try very hard to imagine that these *particular* bears would be *particularly* anxious about the presence of technology—after all they had been tranquilized and operated on and surgically implanted with devices that broadcast their biometrics and location. They had been quantified.

Marshall McLuhan was fond of saying that technology turns the human body inside out, but if that is true for humans it is certainly also true in this instance for the bears.[4] Naturally, the scientists received standard ethical approval for the study (from the University of Minnesota Institutional Animal Use and Care Committee) and this technological inside-out is not quite the same as the more literal laboratory dissections or other experiments that also accompany the pursuit of science. But these factors all seem fair game to an analysis looking to study the impact of human technology on this specific group of bears, even if the bears themselves are, for the scientists, seemingly more metaphors for impact than actual animals of concern. And perhaps a first lesson to be learned from the study is that this is just what a "test subject" is—a metaphor. Each bear is both itself and an instance of experimental change, measurements and data. The bears and their data are incompossible and yet the (scientific) claim is that the data somehow represents the experience of the bears.

However, if, as Boetzkes suggests, "an ecological stance involves revealing the limits of an anthropocentric worldview and recognizing these limits as thresholds to the excess of the earth,"[5] then one thing that is clear in this instance is that there is no real recognition of limits by Ditmer and his team. Instead, the scientific ear attunes to the broadcast data, but in the process becomes oblivious to its own presence even while supposedly benign forms of intervention (abduction and surgical implantation) form the conditions of data generation in the first place. A textbook case of anthropocentric conceit.

To follow Boetzkes's advice in an attempt to think through the complexities of Ditmer's study would be to first ask where the limits of the anthropocentric reside in this particular case. That is, at what point does the study create its own conditions of possibility? For Boetzkes this is a technological question, one she (re)reads through Martin Heidegger's theory of technological limits. What Heidegger described as the tool revealing its essence—"the broken tool springs into view when its dysfunction jars it out of the seamless order of equipment operations"—for Boetzkes is the ecological condition of technological implementation.[6]

This is not simply a technical point; it is also one of the tenets of technological and aesthetic culture. McLuhan might have put it best when he claimed that, pushed to a limit scenario, technology will always reverse into its opposite.[7] The medium is the message, but the message is articulated most strongly when the medium reveals its expressive or technological limits, often through failure. Paul Virilio makes a similar point when he insists that the accident is the hidden condition of any new technology.[8] Boetzkes goes a step further however to insist that at this point the ethical stakes of a technology reveal themselves as well. Following Luce Irigaray, Boezkes calls this "recessive ethics," the act of stepping back from the drive to explain or measure—or of having one's movement towards an explanation conscientiously interrupted—and in so doing also initiating "a retraction from the mistaken presumption that one knows the other, in order to let the other present itself on its own terms."[9]

To look back to the black bears with this in mind is to move from technological generation of data to understanding the study as a contextual and environmental intervention, looking not at what Ditmer has discovered but at what he has created. It is not just a metaphor; it is a metaphor that fails the only way a metaphor can—by poetically charting its own limits of representation. The failure of a metaphor will not simply articulate the (failed) conflation that was the scientific intention, but also the differences between the various components. A failed metaphor, in its failure to constitute intelligible sameness, becomes a poignant articulation of difference. And so, a question: What happens when the weird aesthetics of technological experimentation are given analytic priority over the scientific pursuit of information—when the data metaphor fails to eclipse the bizarrely choreographed performance of scientists, drones and cyborg bears? That is, what happens when science begins to use artistic methods? Is it possible that what seems like science fiction is really just a harbinger for the end of science itself and the birth of something entirely different: the future as an art performance?

The Trouble with (among other things) Jellyfish

At the 2015 meeting of the Society for Literature, Science, and the Arts, the artist Mark Dion gave a keynote lecture in which he spoke about, among other things, one of his installations called *The Trouble with Jellyfish*.[10] The installation was a taxonomy of scientific, natural, and historical representations of jellyfish—framed as many socially-oriented projects are towards a doubled aesthetic and educational mandate. The installation even included an aquarium filled with real jellyfish. They were not wired up and monitored by scientists, but they were put on display for public aesthetic appreciation. Also on display were informative panels with scientific data, artistic renderings, and works produced by local students. But it was something Dion mentioned as almost an aside in his talk that captured my imagination, something that I thought brilliantly recasts the way we conceive of our relationship to the planet and the environment. In the talk, Dion recounted the facts of accelerating climate change, global warming, resource depletion, and overpopulation that define the age of the Anthropocene as one of global human impact, noting that as we impact the planet we are quite literally ruining the world for ourselves. He also said something else, that, strangely, these changes are not equally terrible for all species. As it turns out, while we might be sealing our own fate by impacting the planet so aggressively, in the process we are actively creating a world that is increasingly conducive to the flourishing of jellyfish.

This is almost certainly not what Joseph Beuys had in mind when he so eloquently spoke of how "everyone is an artist" and together we are creating the "total artwork of the future social order."[11] What would happen, however, if we entertained precisely this possibility: that we are indeed living in what, for Beuys, would have been the future, and collectively this is the future we have made? If the Anthropocene is to be defined by the geological impact of humans on the planet, does this not in fact mean that we are responsible for the current state of the world, the perversion of the planet into an unintentional human creation—even, from some perspectives, an artwork of geological scale? It does not mean that it is necessarily a good artwork, but, nonetheless, there may be something to gain by considering it in this way. Less as a trajectory towards the nihilism of human self-destruction and more a species transition project, reshaping the planet itself (even if inadvertently) such that jellyfish are recast as our survival species. Imagine the future of the world we are creating—without us perhaps—but filled with the translucent dancing undulations of underwater interaction: a final choreography enabled by the art project of the Anthropocene itself. Boetzkes writes:

> Beuys took the view that everyone should participate in the shaping of society, and in so doing every person could realize his or her potential to be an

> artist. This vision of society as sculpture, as itself an artwork that was ever in the process of being shaped by the public, also aimed to reintegrate nature into the human social structure. Otherwise put, Beuys's energy plan was to redefine art as a public dialogue that included all citizens, as well as animals and elemental forces.[12]

To think about the anecdote of the black bears as a component part of the artwork of the present is to suggest that this scientific experiment does not just *seem* like an artwork but actually *is* one. Or, at least, an experiment of our own might be proposed in which Ditmer's study is considered an artwork of the present such as to allow for a different kind of impact analysis. For, as Boetzkes puts it, "art has a part to play in critiquing the ways we frame nature through representation as well as through science and technology. Moreover, it does so by forging an aesthetic awareness of how nature exceeds these discourses and representations."[13]

This story really does have all the components of a gripping science fiction novel—or even better as a piece of contemporary performance art—particularly when read through the lens of what is often referred to as the "social turn" in contemporary art. While optimistically defined by Nicolas Bourriaud's theory of "relational aesthetics" (defined as a form of art that uses social space as its medium[14]), it is Claire Bishop that most poignantly counterpoints Bourriaud's vision with a series of examples designed to highlight the perverse, exploitative and sometimes violent forms that such a socially-oriented aesthetic approach can take. Bishop borrows the term "artificial hells" from André

Breton to describe this form of work, as a nod towards the surrealist appeal for "more bold, affective and troubling forms of participatory art and criticism." Such forms often involve what she terms "delegated performance" in which other people—or perhaps, in this case, animals—are set up for a certain type of relational experiment.[15] Importantly, for Bishop the "social turn" in art is only made critically intelligible by importing theoretical concepts from political philosophy—the historical tools of aesthetic analysis are, in her view, not sufficient for the task of parsing social and political complexities of this sort.[16] One might thus speculate conversely as well and suggest that when science begins to behave in strange ways that remind us of fiction and performance art it may be time to think about an *aesthetic turn* in science (or in the collective living of human life) as a way to understand some of the strangeness of contemporary technoculture.

What is it Like to Be a Coyote?

This is not simply the object-oriented proclamation that "aesthetics becomes first philosophy."[17] It is actually the insistence that while aesthetics might be the first philosophy it will also almost certainly be the last. The age of the Anthropocene is not only an age of human impact; it is the age of fatal aesthetic impact, inadvertent human customization of the planetary landscape. This is a performative collapse of the boundaries between art and life, but it is also one that takes seriously the insistence that performance is not a rhetorical category but one infused with embodied and phenomenological consequences. Instead, this foreclosure of critical distance guarantees that—when seen aesthetically—the story of the bears is also one that implicates us in its contextual matrix, whether fused by questions of politics, ecology, or technological living. To accept responsibility for ecological crisis as an "artificial hell" is to acknowledge that each of us had a hand in (collectively) shaping the form in which our present world manifests. Boetzkes calls this "ecologicity," the act of "attuning vision to an ecological reality" that cannot be predicated on a vision of nature that is isolated from human impact.[18]

What is at stake in this inquiry is the question of how we communicate with the world around us, as well as the responsibility we accept for our part in creating the context of such interactions. To think scientifically is to insist on a language of statistics, the conceit of which is to pretend that numbers transcend human bias and, as such, gesture towards a non-anthropocentric mode of engagement. I doubt the black bears would agree however, and, at the risk of anthropomorphizing, it seems reasonable to entertain the possibility of giving these animals a voice in the question. The question is, how?

One approach to the question might be to ask how to attempt to reframe the analysis of the experiment such as to try and understand what it means *for the bears*? This is what the scientists purport to be doing, but it is not what they do. To understand how data registers a calculated intervention is not the same thing as to attend to the perspective of a bear. In fact, following thinkers such as Thomas Nagel or Ian Bogost, it would be tempting to say that the question of what it is like to be a bear is one that is exactly unanswerable, even impossible, for the simple reason that we are not bears. Indeed, for Bogost, "we can never understand the alien [or the bear] experience, we only ever reach for it metaphorically." [19] The fact that the metaphor inevitably fails is not (for Bogost, or for us) an argument against the attempt as much as it is an acknowledgment of precisely the sort of limit scenario Boetzkes alludes to.

Steven Shaviro, for instance, suggests that "a certain cautious anthropomorphism is necessary to avoid anthropocentrism," which I take as a first step towards trying to understand a perspective that is neither simply a projection of oneself, nor a forced reduction to feigned (data) impartiality.[20] Perhaps we should once more look back to Beuys, for just as he suggested that we are creating the artworks of the future (and consequently the artwork of our present), so too did he himself propose an artwork that was generative of the kind of future he wanted to imagine. Take for instance that moment when Beuys locked himself in a cage with a wild coyote for 3 days, a piece in a performance called *Coyote, I Like America and America Likes Me* (1974). A comparison to Ditmer's black bears experiment is in some ways useful here, since the coyote used in this performance can surely be cast as another subjugated animal put into the service of a human spectacle. At the same time, however, there is a proximity that Beuys creates between himself and the animal, not a disavowal of impact but a foregrounding of the relationship proposal with all of its complex interpretive possibilities. Beuys cages himself too—for instance—and spends the next three days making (proverbial) friends with the coyote. Boetzkes insists that "far from attempting to tame the animal, Beuys was instead attempting to discover a form of communication in spite of the animal's wildness."[21] This is not without problems however, since the coyote, while said to be wild was—according to *New York Magazine*—actually procured from a local animal trainer.[22] At the same time one might at least propose the performance as another example of an attempt—even if, perhaps, another "failed attempt"—to create alternate scenarios in which non-human-centric forms of communication might emerge. Beuys called works like this a "symbolic beginning."[23] But he might also have called it the beginning of a conversation in which (some) humans express their intention to live in a world where coyotes (or other animals) also flourish.

What is central is not simply that it is possible to look at our present and future as an artwork in development but that artworks themselves are generative of a certain philosophical—and sometimes ecological—relationship.

Ecologies of (Imaginary) Friendship

What do jellyfish, a coyote, and a family of black bears have in common? Maybe nothing, but there is a component of Beuys's work that gestures towards something that Donna Haraway has turned into a central tenet of her theory of "significant otherness." At minimum there is a certain imagining of possible relationships in play—possible forms of ecologically-conscientious relationships that, while not without problems, begin to form a series of gestures towards framing ecological interactions differently. If, as Haraway argues, "the relation is the smallest unit of measurement"[24] then what we inevitably share with each of these ecological entities is at least a series of possible relations. Yet how the relationship is negotiated matters, as do the ways we position ourselves in their formation. And importantly, for Haraway, such relationships cannot pre-exist their negotiation: "The question turns out not to be what are animal rights, as if they existed preformed to be uncovered, but how may a human enter into a rights relationship with an animal?"[25]

Haraway's suggestion is to think about models of family and kinship as starting points for consideration, with a caveat that over-identification with animals risks personifying, infantilizing, and ultimately compromising the relationships we build. This is one reason she resists the notion of friendship as a valuable form of interaction, since it risks over-identification (for instance, a dog owner who thinks of his or her pet as a personal friend). [26] However, it is also possible that the real problem is not one of over-identification but of under-sympathizing with entities outside of direct human reference. Haraway may critique the dog owner for over identifying with his or her pet as a "friend"—just as I might object to the scientist thinking about a bear as a test subject—but the fact remains that these are both possible forms of relationship-building, though their ethical and ecological consequences are different.

It is possible that to get closer to the story it is necessary to *imagine*, for indeed as Haraway argues, "stories are much bigger than ideologies."[27] To then approach the story of the black bears in this way is to realize that we have much more in common with these animals than we might have first assumed. For these black bears are actually bears, but they are also allegories for our own social, political, and aesthetic situations. If someone tranquilized me, implanted me with a cardiac monitoring system, and then set me "free" to be monitored at a distance, I would certainly be a bit nervous when I noticed the sounds and signs of my abductors circling around me. In so many ways my human situation is really not that different—my running watch came with a heart rate monitor that my phone can read, my phone can automatically upload those biometrics to the cloud, and my data is monitored by algorithms of all sorts to the point where I am sent activity assessment reports and even suggestions for how to upgrade to newer devices that will further improve my performance. Am I a metaphor for the bear, or it for me? Unlike the bears, I suppose I have a choice. Unlike the bears, I grow to love and depend on the suggestions I receive for my donations of data. Unlike the bears I am ostensibly optimized by this relationship—not simply a choreographed test subject, though I'm not really sure it is all that different. If I think of it as a performance it makes me feel a little better—but to be honest it perhaps feels a bit more like science as perverse pedagogy. My sympathies are with the bears, but in many ways my reality and that of these animals may be closer than it first appears.

I am also interested in the idea of *possible* relationships, not yet bound to particularities or subject to the interpretive projections we mobilize in their formation. Because there are no pre-established rules regarding how we speculate on these questions, I think there is a certain merit to considering "friendship" as a model of engagement to add to the mix. As psychologist William Rawlins notes, friendship is not bound to institutional definitions the way that other forms of relationships tend to be. According to Rawlins, family relationships as well as romantic commitments both involve certain formalized behaviors and conventions that are as much socially mediated as they are inter-personally determined. Friendships, for Rawlins, do not have the same kind of social and cultural duties associated with them, which means that the terms of engagement are almost exclusively built and negotiated on the fly between the friends themselves, often resulting in "contradictory terms of engagement."[28] In friendships, Rawlins notes, behaviors may be erratic and sometimes even hurtful, but the friendships persist because of an intent to preserve an "assumption of good intentions" upon which future possibilities for engagement depend. In many ways, though he doesn't use these words, Rawlins's theory assumes a performative trust at the core of friendship—not a contract-based or technically mediated form of relationship, but one built on willingness, suspension, and belief. In fact, it does not take too much imagination to see a link between Rawlins's idea of an "assumption of good intentions" and Samuel Taylor Coleridge's 19th century notion of aesthetics as the *request* for a "suspension of disbelief" in the act of poetic (or in this case, perhaps ecological) faith.

To re-imagine possible relationships in this way is not simply to create new forms of friendship; it is to create possible forms of imaginary friendship. To think of cyborg black bears as imaginary friends is not necessarily an intuitive suggestion, though it is one that animates the impact of a larger culture too often reduced to logistics. To anthropomorphize in this way is not so much about projecting agency onto the bears as it is about acknowledging the limits of our own capacity to dictate the terms of the relationship and to understand how a friendship—imaginary or not—might be a different way of acknowledging the limits and conditions of the relational stage. Boetzkes calls this "a commitment to others beyond anthropocentric boundaries."[29] I might call it an ecology of (imaginary) friendship.

Notes

1 Amanda Boetzkes, *The Ethics of Earth Art* (Minneapolis: University of Minnesota Press, 2010), 16.
2 Mark Ditmer et al. "Bears Show a Physiological but Limited Behavioral Response to Unmanned Aerial Vehicles." *Current Biology* 25.17 (August 2015) 2278-2283.
3 Amanita Khan. "Don't like drones in your backyard? Neither do black bears, study finds." *Los Angeles Times*. August 14, 2015.
4 Marshall McLuhan, *Understanding Media: The Extensions of Man* (New York: McGraw-Hill, 1964), 38.
5 Boetzkes, *The Ethics of Earth Art*, 3.
6 Amanda Boetzkes, "Interpretation and the Affordance of Things," in Amanda Boetzkes and Aron Vinegar, eds. *Heidegger and the Work of Art History* (Burlington, VT: Ashgate Press, 2014), 272.
7 McLuhan, *Understanding Media*, 38.
8 Paul Virilio. *Politics of the Very Worst: An Interview by Philippe Petit*. Sylvère Lotringer, ed., Michael Cavaliere, trans. (New York: Semiotext(e), 1999), 89.
9 Boetzkes, *The Ethics of Earth Art,* 62.
10 Mark Dion. Keynote talk at *After Biopolitics: Society for Literature, Science and the Arts 2015*. Rice University, Houston, TX. November 12, 2015.
11 Laurie Rojas, "Beuys' Concept of Social Sculpture and Relational Art Practices Today," *Chicago Art Magazine*, November 29, 2010.
12 Boetzkes, *The Ethics of Earth Art*, 36.
13 Ibid., 2.
14 Nicolas Bourriaud, *Relational Aesthetics* (Paris: Les presses du réel, 1998), 13.
15 Claire Bishop, *Artificial Hells: Participatory Art and the Politics of Spectatorship* (London: Verso, 2012), 6.
16 Ibid, 12-13.
17 Graham Harman, cited in Steven Shaviro, *The Universe of Things: On Speculative Realism* (Minneapolis: University of Minnesota Press, 2014), 13.

18 Amanda Boetzkes. "Ecologicity, Vision and the Neurological System," *Art in the Anthropocene: Encounters Among Politics, Aesthetics, Environments and Epistemologies*. Eds. Heather Davis and Etienne Turpin (Ann Arbor: Open Humanities Press, 2015), 276
19 Ian Bogost, *Alien Phenomenology or What It's Like to Be a Thing* (Minneapolis: University of Minnesota Press, 2012), 66.
20 Shaviro, *The Universe of Things*, 61.
21 Boetzkes, *The Ethics of Earth Art*, 37.
22 Thomas B. Hess, "The Germans are Coming! The Germans are Coming," *New York Magazine*, June 24, 1974, 54.
23 Johannes Stüttgen, *Beschreibung eines Kunstwerkes ("Description of an artwork")*, (Düsseldorf: Free International University, 1982), 1.
24 Donna Haraway, *The Companion Species Manifesto* (Chicago: Prickly Paradigm Press, 2003), 24.
25 Ibid., 53.
26 Ibid., 94-96.
27 Ibid., 17.
28 William Rawlins. *Friendship Matters: Communication, Dialectics and the Life Course* (New York: Aldine de Gruyter, 1992), 9.
29 Boetzkes, *The Ethics of Earth Art*, 182.

6 Feeling with Your Eyes
Petroculture and *Ice Follies*

Serena Kataoka

Here we are. In worldwide webs of plastic objects, whose "plasticity" seems to be foreclosed by global oil. In a world laid bare such that *revealing* the truth of oil disasters is transmuted into possibilities of and for profit. In an aesthetic worlding of a plastic condition in which *dissensus* maps (rather than remaps) the sensible, rendering criticality sterile. It may be tempting to imagine an exit strategy from this "*exploding plastic inevitable*: the predetermination of a future oil regime and its reification through a mesh of plastic objects."[1] But in this catalytical textual space, I take seriously Amanda Boetzkes's claim that, "Petroculture is lived from within."[2] The question she then poses is, "What kind of vision can recover us from plasticity and plasticity from us?"[3] Given the dominance of dualistic thinking (we *subjects* versus those plastic *objects*), how might we come to see that we and plasticity are *one*, and yet different—in a way that respects the force of things (the ability of plastic objects to obscure and reveal petroculture)—so that we can come to see plasticity and ourselves otherwise?

What follows is an experiment in seeing plasticity otherwise. It is grounded in ways of living and a place that I am familiar with, by virtue of *contingency* (as distinct from agency or accident).[4] While "I" may seem to dominate the analysis, this subject-centrism is largely an inherited problem of the common use of English, within a tradition of privileging civilized subjects. I am not special. I do not have any magical powers with which to divine or otherwise determine the future through action. So while the first section of this chapter will note that I was born and raised in a Buddhist community, it will also gesture toward other *conditions* of possibility (or virtual reality) of coming to appreciate oneness, such as Indigenous ways of living and philosophical monism.[5] Likewise, the second section will encounter petroculture on Nipissing territory, on land that I occupy, in the messiness of living colonialism and with commitment to improving our ways of living together here. But one might encounter petroculture in virtually any place, these days. So readers may find it creatively productive to think in and through other conditions and places.

This chapter is a provocation to experiment with coming to another way of seeing petroculture. As fodder for doing so, I offer some vignettes from *Ice Follies*—a biannual festival of site-specific art on frozen Lake Nipissing, whose 2016 theme explicitly addressed global oil and its relations to water. It is my hope that whatever affects these vignettes might have, and whatever affinities might be constructed through them, will exceed common sense and help to unravel reigning imaginations of plasticity.

Coming to See Otherwise

Seeking a kind of vision—a sight and a way of seeing—raises a question of approach. How can one come to see otherwise? Growing up as a *haafu* (a half-Japanese person) in a Japanese Buddhist community in Canada has informed my sense of "the universe" as having a fundamental principle that informs everyday life. While I necessarily write as someone who was born into that community, I do not write with a consciously Buddhist approach, because for me Buddhism is not only a philosophy. It is a way of living, in which *study* makes sense along with *faith* and *practice*;[6] the force of committing and submitting, and attending to everyday actions cannot be translated well into language.[7] There are certainly others who embrace the challenge of writing through Buddhist ways of thinking,[8] but for me, writing is moved by a broader commitment to improving our ways of living together, and so extends beyond my personal practice and community. While walking a Buddhist path, I happily accept invitations to wander with others along parts of their paths, as we build relations of interdependency and co-operation.

As an undergraduate refugee from western approaches to communities in International Development, I found a welcoming place for learning how to mobilize in and with communities in Indigenous Studies. I continue to work with Indigenous people and communities—with whom I seem to share a similar sense of "the world alive"[9]—but I have not been gifted with particular stories or teachings (and so will not speak to any of those directly). After a couple of years living in Fernie (a small city in Western Canada) and working with community members (who often had conflicting commitments to resource development, environmental protection, and social justice) to improve that community, I attended the University of Victoria to do graduate studies (in the hopes of earning a living wage doing community-based work). There I encountered philosophies of oneness—so-called "monism" (including works by Heidegger, Deleuze, Hegel, and others)—which helped me appreciate the ways that communities are informed by (and can exceed) structural conditions, mythologies and histories; but I always felt like a Buddhist in intellectual drag, trying on various concepts and philosophical moves. I have learned much about improving our ways of living together through both Indigenous ways of living and philosophical monism, but those paths are different than mine. It is no mere accident, however, that I find them to be productive catalysts.

Buddhist, Indigenous, and philosophical monist approaches tend to share concern about the violence inherent in dualistic thinking (subject/object, good/evil, white/racialized, civilized/barbarian, etc.). In addition to carving the world up into myriad battlefields, such binaries overdetermine the "winners" and so underpin imperial impulses. The subject/object binary, for example, sets up one's drive for mastery over some thing, and ultimately every thing.[10] Happily, we are not alone in resisting dualistic thinking. So I hope that readers sharing this concern but walking different paths might read their own approaches into this piece (or use it as a catalyst to chart another one). Allies in the resistance to western imperialism need not have an intimate understanding of one another's paths in order to co-operate, but being-against is not a sufficient basis for creating other ways of living together.[11] More affirmative ways of coming together are in order. But finding, lingering in or amplifying *affinities*[12] among our ways requires that we resist the tendency of our assumptions to overwrite encounters. For instance, I have been trained to see global oil as one among many movements (such as green capitalism, localism, and environmentalism) jockeying for influence in a "global urban" hyperspace of sorts—a political space that we have to imagine in order to make sense of everyday relations.[13] At ease in my analytical tendencies, then, I might become preoccupied with making fine intellectual distinctions, between "space" and "place," for instance.[14] While such distinctions have a place/space, my concern is that they can get in the way of cultivating affinities.

So rather than making fine intellectual cuts, I am interested in doing the inverse: searching for critical methods that allow for blurred, messy or affective forms of collaboration and engagement. I am interested in fuzzy ways of seeing, and so I turn to Carlos Castaneda's mythology. Because there is some doubt as to whether his encounters with a "shaman" actually happened, and because his reflections were informed in large measure by experiments with psychotropic drugs, his work is not to be taken too seriously (most especially as relaying Yaqui culture). The appeal of Castaneda's work, however, is popular (his 8 books have been translated into 17 languages and have sold more than 8 million copies worldwide), and so I read his work as a "New Age" mythology accessible to a largely western audience trying to stretch beyond monotheism grounded in good/evil. There are no doubt other mythologies that one might find productive for making one's sight a little fuzzier, so feel free to insert another one here if Castaneda does not work for you. Otherwise, in seeking cues about how we might linger in and with affinities, we might consider one of several strategies Castaneda offers:

> After a long pause don Juan suddenly turned to me and said that in order to find the proper place to rest all I had to do was to cross my eyes …. Don Juan then described the technique, which he said took

> years to perfect, and which consisted of gradually forcing your eyes to see separately the same image. The lack of image conversion entailed a double perception of the world; this double perception… allowed one the opportunity of judging changes in the surroundings, which the eyes were ordinarily incapable of perceiving.
>
> Don Juan coaxed me to try it. He assured me that it was not injurious to the sight. He said that I should begin by looking in short glances, almost with the corners of my eyes. He pointed to a large bush and showed me how. I had a strange feeling, seeing don Juan's eyes taking incredibly fast glances at the bush. His eyes reminded me of those of a shifty animal that cannot look straight. …
>
> He patiently explained that looking in short glances allowed the eyes to pick out unusual sights.
>
> "Such as what?" I asked.
>
> "*They are not sights proper,* he said. *"They are more like feelings.* If you look at a bush or a tree or a rock where you may like to rest, your eyes can make you feel whether or not that's the best resting place." …
>
> "I don't care what you see … You could be seeing an elephant. How you feel is the important issue." …
>
> "*The trick is to feel with your eyes,*" he said. "*Your problem now is that you don't know what to feel. It'll come to you, though, with practice.*"[15]

Seeing cross-eyed and with darting glances, then, destabilizes the appropriation of a single coherent image, and encourages one to attend to changes over time and ease into a more elemental sense-ability. Trying to see in this way, approaching the imaginary space where Buddhist, Indigenous, monist and other paths criss-cross, where might we rest together? Perhaps we might linger in a sense that every*thing* has affects (whether categorized as a person, artwork, rock, or whatever else), in a sense of co-responsibility (of beings, forces, relations, causes, etc.), and in generous curiosity about those things, affects, and co-makings. Most especially, however, at this intersection, we are redirected toward everyday life, where these sense-abilities are actualized and virtualized.

But too often (and too quickly), when we turn to everyday life we get caught up in focusing on particularities. The challenge, then, is to take the focus on information about and experiences of this place, and hold both visions simultaneously, so as to glean a fuzzier impression.

Focus: Nipissing Territory & Petroculture

Given the ubiquity of global oil, one could presumably encounter it virtually anywhere. But I begin *here*—on Nipissing territory—because it is the land I occupy, and it also happens to be a key node in current struggles over petroculture. The analysis in this section is deceptively focused. Yet, I cross two perspectives: through one eye, I provide a data gaze. I trace geographical, historical, and social markers, creating a *representation* of this place as relatively remote and in decline. By utilitarian capitalist logics, the situation here is relatively dire for all but the affluent few whose privileged mobility transcends locality. Through the other eye, I offer a lived gaze. Gesturing toward Indigenous survival and resurgence, and how residents more generally love it up here, it is possible to create a sense of lived *experiences* of this place. Privileging holism and "quality of life"—the valuing and enjoyment of many things—everyday life here is marked by a sense that "we take care of our own," even (if not especially) those who are struggling. Crossing these two visions (one representational and one experiential), what might you feel?

Nipissing territory is geographically remote by city-centric standards: a three-and-a-half hour drive northward from Toronto or westward from Ottawa, the urban form peters out into tracts of cultivated lands—punctuated by occasional settlements—cutting eventually into and across the Cambrian Shield. Dustings of soil are thin invitations for boreal forest to root into or pry open cracks in the granite rock. From this height of land, water spills from sandy Lake Nipissing westward into Georgian Bay and the Great Lakes, and eastward from the rocky depths of Trout Lake down the Ottawa River shed into the St. Lawrence River, eventually. The recent Energy East proposal to convert the natural gas pipeline running by both lakes into a crude oil pipeline, has made this place a flashpoint in national struggles over petroculture and our ways of living together. But a bit more context is needed to get a sense of the forces at play in these struggles.

The urban hub of Nipissing territory is North Bay. Its survival seems to have hinged on luck, Fortuna, or some other such force at pivotal moments (which is not to downplay the entrepreneurialism and hard work that sustains this place otherwise). Having "missed the boat" as a key node in the Saint Lawrence Seaway (despite the fact that the northerly route was a rational choice, requiring fewer locks, for instance),[16] North Bay became a railway hub and administrative headquarter (of the Canadian Pacific Railway, initially, and more recently the Ontario Northland Transportation Commission). So despite lacking the natural resource industry base of nearby Sudbury, for instance, this place became a regional centre. On the heels of massive layoffs at the railway during the Great Depression, the city was on the brink of collapse when the miraculous birth of the Dionne Quintuplets drew some 3 million tourists through the region, kick-starting economic recovery (from 1936-43).[17] Then, with the Cold War on the horizon, thanks to having won the railway and geography lottery, North Bay was selected as the site for the NORAD's Canadian air defence command and control centre—which is buried some 680 feet underground in a granite hill, where it would be less susceptible to Soviet nuclear strikes.[18] Thanks to these boons in infrastructure, tourism, and employment over the last century or so, North Bay has become a regional hub with a hospital, university and college, provincial government offices and jail, as well as retail depots.

Yet again, though, we are teetering at the edge of decline—what with the relatively recent provincial decision to divest (now part of) the railway company,[19] which is compounded by mass layoffs at the hospital[20] and cutbacks at the university,[21] all of which will eventually impact the retail and services sector (which is one of the two biggest employment sectors in town). There is serious cause for concern given that the city *already* has triple the homelessness rate, per capita, of Toronto.[22] Few residents are optimistic about North Bay's future (19%), and two thirds believe that North Bay is missing key business opportunities.[23] While most residents think they could find similar employment elsewhere in Ontario (64%)—in the event that they lost their current jobs—only 10% think they could do so in North Bay or Nipissing region.[24] Given this dire-seeming picture, why do people stay here, and struggle to make things work in and for this city? I propose that there is more than Fortuna at play.

We love up here.[25] Despite lacking optimism about the city's future, the vast majority of residents (79%) enjoy living in North Bay.[26] The top three best features of North Bay, according to a recent survey of residents, reflect the dual sense in which we love up here. On one hand, we love *up here, i.e.* the place itself—its beautiful views (75%) and outdoor activities (71%). On the other hand, we *love* up here, *i.e.* care for one another—its friendly people (51%).[27] Anecdotally, many folks—myself included—are on Nipissing territory to be close to family and friends. I share a home with (and care for) my brilliant brother, am an hour's drive from my parents in Huntsville, and a three-and-a-half hour drive from other close family members in Toronto, Ottawa, and Peterborough. By day, I work as the Executive Director of the White Water Gallery (a roughly 40-year old artist run centre on Main Street), and by night I teach social justice oriented courses at Nipissing University. While I could have higher earnings and more job security elsewhere, *here* is where I can go for a swim/cross-country ski after work, and squeeze in dinner with my parents or take in a show before unwinding at home with my brother. While other residents' stories are different than mine in their particulars (*e.g.* many might opt for fishing, hunting, snowmobiling, hockey, volleyball, theatre, or a barbeque), there seems to be a common thread of valuing what urbanites might call "quality of life." We are doing more than getting by. This *love up here* is imposed upon, builds on, and sometimes cooperates with a much longer-standing, and deeply-rooted tradition.

Nipissing nation has been thriving at this crucial juncture of canoe routes and pathways for thousands of years—a living fact both confirmed and obscured by a land claim that "settled" boundaries, and provided $123.9 million in financial compensation, most especially for the loss of historical use, inflation, and fair market value of lands that should have been included in the Nipissing reserve (based on the 1850 Robinson Huron Treaty).[28] On "official" maps, Nipissing 10 reserve appears as four small communities scattered around the edges of the City of North Bay, but together they are rooted in Nipissing nation's Chi-Naaknigewin (*i.e.* constitution)—reportedly the first of its kind in Ontario.[29] Whatever Canadian courts might "rule" when the Chi-Naaknigewin is "tested," Nipissing nation is actively self-governing. For instance, in addition to providing services one might expect from a local form of government (*e.g.* recreation, social services and programming, economic development, etc.), Nipissing nation has been offering Indigenous education through Nbisiing Secondary School for over 20 years, and is implementing its own Fisheries Law, based on Indigenous authority and knowledge (which is no less meaningful for having also been affirmed by the provincial Ministry of Natural Resources and Forestry).[30] Nipissing nation's leading role may be seen as part of an Indigenous resurgence movement that is grounded in the land, and is transforming our ways of living together.[31]

So to get a feeling of this place, one could cross these two visions—a representation of decline, and experiences of Indigenous resurgence and loving up here.[32] I cannot tell you what to feel, but imagine that it will be messy. If both visions are accepted as valid in their own ways, but not necessarily in dialectical tension, then there is no reigning thought or emotion.

It is in and with that ambivalent sensibility that we encounter petroculture on Nipissing territory. TransCanada Corporation's proposed conversion of an existing 40-year old natural gas pipeline (from Saskatchewan to Ontario) to carry 1.1 million barrels of tar sands oil per day (from Alberta to a new export port in New Brunswick) cuts right through the Trout Lake and Lake Nipissing watersheds.[33] So people here are rightfully concerned about the "risk" of contamination of the municipal water source,[34] in addition to serious environmental damage; residents are especially sensitive to that prospect, given a formaldehyde spill in Trout Lake from a nearby highway in Spring 2012.[35] So in addition to environmentalist opposition that one might expect (*e.g.* through the local chapter of Stop Energy East), the generally business-oriented mayor of North Bay, Al Macdonald, has taken a stand against the pipeline conversion, seeking intervener status through the National Energy Board.[36] Furthermore, Energy East is bound to consult with First Nations, and Nipissing nation is a proponent of green energy. Nipissing Chief McLeod is concerned about risks to Lake Nipissing through potential contamination of Duchesnay Creek and the Little Sturgeon River, and Liberty-Whiteduck—"the grandmother who has been blessing the waters of the two important lakes nestling North Bay and Nipissing First Nations for many years"—"says it is the duty of all of us to protect the water for seven generations to come."[37] So across communities and the ideological spectrum here, in spite of our heavy reliance on oil in our everyday lives, *this place*—the land and relations of, to and on it—means more than moving dirty oil.

Water has become symbolic of what is threatened by and yet also exceeds petroculture. Creative activist interventions here have included: artist Kim Kitchen's intervention at a TransCanada consultation, where she presented a crystal pitcher full of water and another full of oil, and offered passersby crystal, fishbowl-like glasses inscribed with: "Check one: ❑ Water ❑ Oil"; a recent Resistance Cabaret;[38] and an NDP rally where canoes were launched into Trout Lake.[39] This local attentiveness to the value of water is part of a global movement,[40] and so we might speak of a *glocal*[41] movement for water. Turning to water is a way of sinking into the production of plastics; it takes somewhere in the range of 1.39 to 6 litres of water to produce a 1L plastic bottle of water (depending on whether one tends toward research done by the International Bottled Water Association or environmentalists).[42] Water is, therefore, a condition of possibility (or virtual reality) of our plastic condition.

Feeling with Your Eyes: *Ice Follies*

Ice Follies 2016 was the sixth biannual festival of site-specific art on frozen Lake Nipissing. A snowy white expanse stretches like a canvas toward a distant horizon, or a "white box" blasted open yielding to the massive sky, thus offering artists a specific yet flexible site for working in and through. These vignettes might be thought of as open-ended presentations of other visions of petroculture, interventions designed to engage different ways of looking, conceiving and collaborating. No attempt will be made to lock down meaning in/as the end, but I hope that you might find affinities in and through what follows, feeling with your eyes (and perhaps your willing imagination).

Moved by concern about the Energy East project and respect for the value of water, the organizers of Ice Follies 2016 selected the theme, *Ojichaagobiishin*.[43] Translated as "It reflects in Water," this was a call to action for artists to reflect on water as the giver of life and our responsibility to care for it. Much is lost in translation, though. *Ojichaagobiishin* ultimately makes sense in Anishinaabemowin, which is the language of Nipissing people (among others). As a non-speaker, rather than feign deep understanding, I offer thanks to Aanmitaagzi—an Indigenous driven artist-run community-engaged arts organization on Nipissing territory, and since 2012, a co-organizer of Ice Follies along with the Near North Mobile Media Lab (N2M2L) and White Water Gallery.[44] Thanks to Aanmitaagzi for making the festival relevant to the culture proper to (because properly *of*) this place, more responsive to people of that culture, and for inviting we newcomers to become more familiar with this place and people—through workshopping and performing stories.

***

Ice Follies installation: It is minus 40 degrees Celsius before wind-chill … again. Artists and their assistants have been installing works on Lake Nipissing all week, and tomorrow is opening day. An assistant teeters on a 15-foot ladder, pulls off a mitten to enable her gloved hand to hold a screw in place, and uses her other mitted hand to drill the screw in place; she replaces the mitten, shakes blood back into her hand, and repeats. Ten screws in, the drill's battery freezes, and so she treks back to the festival tent to swap out the battery for a charged and thawed one. 20 screws to go. This action is just part of one installation, and at this glacial-seeming pace it is hard to imagine how everything will get done, but it is amazing what can be done for the love of art (and hot chocolate). The cold and gusts of wind seem to add to the adventure of working on this snowy expanse, but a couple of artists can be found at any given time huddled in a car running in the parking lot with the heat cranked up. Safety first.

***

Edgardo Moreno, *Nisidotan* (Understand by Hearing):[45] Standing at the centre of a circle of speakers on the ice by the town dock, you are surrounded by sounds of water gurgling and lapping, ice creaking and cracking, and wind rustling—all of which become quiet seeming as sounds of a plane (or is it a train?) come from behind and booming through your chest, it seems to fly right overhead as the sounds shift to the speakers in front of you. These curated sounds (gathered from Lake Nipissing over the previous months) interact with the live sounds of footsteps crunching in the snow, children laughing, the hum of cars on the nearby road, wind rustling in the trees, and so on. Immersed you are, but what resonates?

***

Aanmitaagzi, *Serpent People*:[46] The sun has just set, unseen rays casting the sky a deep purple. You are drawn toward a group of about 100 people gathered next to a heated tent, and there you fold into the warmth and excitement about the performance to come. Lights dance on the snow, and a narrator's voice cues you to enter the *Serpent Tale*, transforming this space into a stage of the Manitou Islands. It is to this circle of five small islands at the centre of the lake that Nipissing people had fled, seeking refuge from raiding parties. Ushered toward a teepee whose covering is woven from birch bark and woolen blankets, you see three dancers hunched over—as starvation began to set in, along with winter. Luckily, it seemed, a child gathering water from a small lake on the island saw *Makadewaa Namewag* (a great black sturgeon) under the ice, and so the people caught the fish. Ravenously, the dancers raise fistfuls of imaginary fishmeat to their mouths, hardly swallowing before gnawing at another piece. But as the meal settles in, smooth scaly material creeps up their bodies, and they are transformed into serpents. You follow their slithering forms across the snow and are ushered into a spiral maze whose 8-foot high fabric walls are lined with images and phrases from community workshops of *Serpent Tale* over the previous months. Sounds of hissing and slithering can be heard from all directions, and it is as though you and the line of people before and after you have come to form serpents yourselves; winding inwards, you eventually reach a hole in the ice at the centre, where the sturgeon had been caught, and into which the serpent people slithered. After winding your way back out of the spiral, traditional drumming and singing fills the air, and you learn that no one has since tried to live on the Manitou Islands.

***

Jihee Min, *A Humble Trowling*:[47] In the stillness of early morning the snow squeaks, more than crunches, as you walk toward a giant fishnet suspended above the lake. A grayscale image of hands cupped together and reaching outward is meticulously woven into the net—a translation of a digital photographic image, with each square marking a pixel. Tails of the thread woven through the net are moved by the wind, and so dance toward you without getting closer. Hands, the symbol of work in a capitalist system, here appear to be scooping up the water of Lake Nipissing, and gracefully offering it up to the sky. The rising sun dusts pink and purple across the horizon, throwing the image into relief.

***

Meg Paulin, *The Trapper's Cabin*:[48] An aged gas pump is mounted on the ice. At one time it was actively used at a gas station on Nipissing nation, but here the pump seems ready to draw water from the lake. Alongside the pump is a small cabin where you can come in from the cold. Referencing trappers' intimate knowledge of the land and tradition, the walls are lined with historical maps, articles, images, and objects—fragments of land use knowledge and local stories gathered through research with Kinnickinick Consulting and Nipissing nation, and community engaged workshops with Aanmitaagzi. You are invited to draw or write your stories, and to sketch pathways you tend to take on a hand drawn map of Lake Nipissing whose green shorelines fade into newsprint. Together, old and new reflections on relations of and to the land—including tensions arising from western impositions—form layer upon fluttering layer of community knowledge.

***

Darren Copeland, *Fishing for Sounds*:[49] A group of eight people venture out on the Ice Follies site, armed with contact microphones that they made on the first day of this community sound workshop, to go *Fishing for Sounds*. Slipping headphones over their toques, participants begin wandering around seeking sounds that tend to be inaudible, looking something like doctors with stethoscopes outstretched. The surrounding sounds on the lake become muted, and so you can hear them calling to one another in louder-than-typical voices, flagging interesting sounds found in different spots on the site. In the vibrations of the posts holding up the fabric walls of the spiral, one participant has found deep, bassy rhythms. Lingering on the ice itself, another participant is surprised to discover virtual silence, punctuated only occasionally by echoes of footsteps in the snow. Meanwhile, workshop organizers have been diligently hammering away at the ice that has sealed up the hole at the centre of the spiral, in the hopes of being able to drop a hydrophone right into the lake. As it turns out, those smashing, cracking, and scraping sounds turned out to be as rich as those that were eventually captured in the still water beneath. What makes a sound a good "catch"?

***

STAG (Student Temporary Art Gallery):[50] Off the edge of the town dock, beyond the installation site, ice-fishing huts are strewn as though each were erected where a stone once skipped—randomly and yet with a visual rhythm of sorts. But these temporary communities actually follow the contours of the lake bottom (above the slopes where walleye/pickerel and others are thought to linger), and so looking at these scatterings of huts is to look at a mirror of the slopes of the lake. There is one hut, though, in the shallows of the *Ice Follies* site that stands alone. Approaching it, you notice that the wall closest to you is covered in a minimalist, whimsical drawing of a young woman. Peering in the window, you see a salon style exhibition of drawing, painting and print work, along similar minimalist and whimsical lines. As you walk back toward the dock, you wonder if there was indeed crocheted lace on the windowsill, or if your imagination has worked that object out of a lingering sense of the micro exhibition as informed by retrospection and care for the work of hands.

***

Caitlind r.c. Brown & Wayne Garrett, *The Deep Dark: Crossing*.[51] Crunching snow underfoot punctuates the seeming silence. Darkness folds around you so completely that there is no apparent horizon. But a series of magical-seeming doorways of white light hover on the snow before you, inviting you to hike out on the lake. As you pass through the first doorway, you are blinded by the light, so on the other side you pause and blink until your vision returns. The darkness seems thicker now, and yet the second doorway is straight ahead, framing the slightly smaller doorway beyond it, which frames the one after it, and so on, into the distance. You have found yourself in a tight-seeming hallway with no walls. Compelled to continue forward, you notice it becomes easier to find footing, but each doorway seems farther away than it had looked from the last one. Finally, after well over a kilometre, you walk through the last doorway and find a bench. Sitting here, your awareness opens up to the expansiveness of what lies beyond the human-scale hallway: the lake gurgling below, the sky and stars, wind breathing on trees, and ....

***

Ice Follies: The wind has been swirling around the installations for two weeks now, making them feel even more a part of the site. Small banks of snow and crystals of ice have begun forming at the base of the fishing net. Many of the children's drawings of serpent people have been blown off the spiral's fabric walls, and then tucked under the wooden posts by audience members—forming something of a spontaneous collage every four to eight feet along. Similarly, some of the mirrored strips on the doorways have peeled off, but passers-by have used whatever they had on hand to put them back together. The tepee, in particular, looks as though it has always been here. Carefully woven strips of woollen blankets—including a Hudson's Bay blanket—hung above the birch bark base have begun to disentangle and strips are blowing about; it is now difficult to decipher the words written on the material by community members, but this seems to make this work of many hands hang together as one.

Together, when we reflect on water, we see it and our mirrored image—both of which are bound up in global oil and other globalizing forms, such as colonialism, in complex ways. Thousands of people gave audience to Ice Follies, many of whom were moved by Indigenous resurgence and/or loving up here. How might you see water and oil, and ourselves otherwise? Cross your eyes, perhaps ... as each of the artists might have done themselves.

Notes

1 Amanda Boetzkes, "Plastic Vision and the Sight of Petroculture," *Petrocultures: Oil, Energy and Culture*, Sheena Wilson and Imre Szeman, eds. (Montreal and Kingston: McGill-Queen's Press, 2017), 23-24.

2 Boetzkes, "Plastic Vision," 7.

3 Boetzkes, "Plastic Vision," 25.

4 *c.f.* Judith Butler, "Contingent Foundations: Feminism and the Question of 'Postmodernism'" In *Feminist Contentions: A Philosophical Exchange*, Seyal Benhabib, Judith Butler, Drucilla Cornell, and Nancy Fraser, eds. (New York: Routledge, 1995), 35-58.

5 On conditions of possibility see Foucault on "episteme." Michel Foucalt, *The Order of Things,* Alan Sheridan, ed. (New York: Vintage, 1973[1966]). On conditions of creative production, of the virtual (as real), see Gilles Deleuze, *Difference and Repetition,* Paul Patton, trans. (New York: Columbia University, 1994[1968]).

6 On "Faith, Practice, and Study" see SGI Canada, *Buddhism: A Philosophy for Living* (Toronto: SGI Canada, 2004), 11-12.

7 *c.f.* practical knowledge. See for example: James C. Scott, *Seeing Like a State* (New Haven and London: Yale University, 1988).

8 Some influential examples include: Daisetz Teitaro Suzuki, *An Introduction to Zen Buddhism* (New York: Grove Press, 1994[1934]). Alan W. Watts, *The Way of Zen* (New York: Knopf Doubleday, 2011[1957]).

9 Dr. José Barreiro, *The Ka Na Ta Conversations* (oral presentation as part of a lecture series co-hosted by Nipissing University and the Assembly of First Nations, April 4, 2013).

10 Martin Heidegger, *Nietzsche: The Will to Power as Art,* David Farrell Krell, ed. and trans. (New York: Harper & Row, 1979), 78.

11 Being-against patriarchy, for instance, has been shown to be insufficient for building a feminist movement that

is inclusive, let alone just, when racism is pervasive. The second wave of feminism, in particular, privileged middle class white women. bell hooks, *Feminist Theory: From Margin to Centre* (London: Pluto Press, 2000). So while this chapter emphasizes affirmative connections between engaged communities, it is with the understanding that building those relations requires that we—especially those of us with relative affluence and influence—do the critical self-reflection necessary to take responsibility for the ways that we each participate in oppressive practices. This is not a matter of identity politics. It is a condition of seeking justice together.

12 *c.f.* Johann Wolfgang von Goethe, *Elective Affinities*, R.J. Hollingdale, trans. (London and New York: Penguin, 1971[1809]).

13 Warren Magnusson, "Globalization, Movements, and the Decentred State" In *Local Self-Government and the Right to the City* (Montreal and Kingston: McGill-Queens University, 2015), 125-155.

14 *c.f.* Massey's critique of Heidegger's conception of place, and her "slices through space." Doreen Massey, "Reorientations," In *For Space* (London, Thousand Oaks and New Delhi: Sage, 2005), 105-146 (See especially, note 17).

15 Emphases added. Carlos Castaneda, *Journey to Ixtlan: The Lessons of Don Juan* (New York: Washington Square, 1991[1972]), 49-52.

16 Backroads Bill Steer, "The Ontario that Almost Was: The French River and the Georgian Bay Ship Canal" *Northern Ontario Travel*, February 10, 2015. http://www.northernontario.travel/northeastern-ontario/french-river-georgian-bay-ship-canal

17 Françoise Noël, *Family and Community Life in Northeastern Ontario: The Interwar Years* (Kingston and Montreal: McGill-Queen's University, 2009), 240-42. Also Françoise Noël, "Old Home Week Invitations and Ephemera 1935" Nipissing University Archives, January 22, 2013. http://faculty.nipissingu.ca/noel/archives/tag/dionne-quintuplets.

18 Bruce Ricketts, "NORAD North Bay: The Story of Living Underground," *Mysteries of Canada*, July 23, 2016. http://www.mysteriesofcanada.com/military/norad-north-bay/.

19 Liz Cowan, "Northerners Fight Divestiture of ONTC," *Northern Ontario Business*, January 5, 2016. http://www.northernontariobusiness.com/Industry-News/transportation/Northerners-fight-divestiture-of-ONTC.aspx. Also Bonnie Lysyk, *Divestment of Ontario Northland Transportation Commission: Special Report* (Toronto: Office of the Auditor General of Ontario, December 2013), Gord Young, "ONTC Divestment would cost $820 million—Auditor Says," *The Nugget*, December 10, 2013. http://www.nugget.ca/2013/12/10/ontc-divestment-would-cost-820-million--auditor, and Jennifer Hamilton-McCharles, "Wynne Commits to ONTC Future," *The Nugget*, August 12, 2016. http://www.nugget.ca/2016/08/12/wynne-commits-to-ontc-future.

20 See for example: Gord Young, "North Harder Hit by Hospital Cuts," *The Nugget*, July 8, 2015. http://www.mysteriesofcanada.com/military/norad-north-bay/.

21 Jeff Turl, "Nipissing University Layoffs Hit 14," *BayToday*, November 27, 2014. https://www.baytoday.ca/local-news/nipissing-university-layoffs-hit-14-21423. Also P.J. Wilson, "Cuts Continue at Nipissing," *The Nugget*, February 5, 2015. http://www.nugget.ca/2015/02/05/cuts-continue-at-nipissing.

22 Based on statistics from "Community Profiles," The Homeless Hub, May 2016. http://homelesshub.ca/CommunityProfiles.

23 Baylor MBA Team, *North Bay: Phase One Presentation* (North Bay: Baylor Business, and City of North Bay, 2016). http://www.investinnorthbay.ca/media/1548/north-bay-final-presentation-phase-1.pdf

24 Baylor MBA Team, *North Bay*, 16.

25 This phrase was directly inspired by the "We Live Up Here" multimedia project to revitalize Sudbury through the arts. "We Live Up Here." http://www.weliveuphere.com/.

My slight, playful twist on their phrase is intended as an honouring gesture of solidarity.

26 Baylor MBA Team, *North Bay*, 12.

27 Baylor MBA Team, *North Bay*, 20.

28 Jennifer Hamilton-McCharles, "Land Claim Settlement Reached," *The Nugget*, March 24, 2013. http://www.nugget.ca/2013/03/24/land-claim-settlement-reached. Also "Land Claim Information," Nipissing First Nation, updated November, 2015. http://www.nfn.ca/land_claim.php

29 "Nipissing First Nation Passes First Ontario Aboriginal Constitution," CBC News, January 21, 2014. http://www.cbc.ca/news/canada/sudbury/nipissing-first-nation-passes-first-ontario-aboriginal-constitution-1.2505488.

30 "Ontario Recognizes Nipissing First Nation's Fisheries Law," *Anishinabek News*, March 24, 2016. http://anishinabeknews.ca/2016/03/24/ontario-recognizes-nipissing-first-nations-fisheries-law/

31 See for example, Leanne Simpson, ed. *Lighting the Eighth Fire: The Liberation, Resurgence, and Protection of Indigenous Nations* (Winnipeg: Arbiter Ring, 2008).

32 Emphasizing affirmative movements of (and so too potential affinities between) Indigenous people and Settlers on Nipissing territory is not to downplay the persistence of colonial violence here. Ample evidence of racist entitlement and fear of losing that which has been taken can be found in everyday interactions (*e.g.* where Settler business people speak of "hitching their wagons" together), as well as in public venues (*e.g.* Bill Eden, "Ministry, Nipissing First Nation Agreement could Spell Trouble for Anglers," *Almaguin News*, April 23, 2016); among the most blatant and offensive statements can be found in the "Comments" section of news articles, and Facebook posts (whose violence I will not replicate here). So much to say that much critical decolonizing work remains to be done by and with Settlers, that demystifies the "Indian problem," by focusing on Settler racism and colonialism. See for example, Adam J. Barker, *Settler: Identity and Colonialism in 21st Century*

Canada (Black Point NS: Fernwood, 2015). By emphasizing potential affinities, this chapter "run[s] the risk of reifying (and possibly replicating) settler colonial as well as other modes of domination." Corey Snelgrove, Rita Dhamoon, and Jeff Corntassel, "Unsettling Settler Colonialism: The Discourse and Politics of Settlers and Solidarity with Indigenous Nations," *Decolonization,* 3(2), 2014, 1-32. And it is a risk I am willing to take (as a Settler), if only as an opportunity to continue to learn to listen otherwise (which is not to justify or minimize whatever injuries and frustrations may be incurred by readers).

33 *When Oil Meets Water: How the Energy East Pipeline Threatens North Bay Watersheds* (Ottawa: Council of Canadians; North Bay: Stop Energy East, April 25, 2016).

34 There are likely to be 6 major ruptures along the pipeline over the 40 years of the project. *When Oil Meets Water*, 3.

35 Maria Calabrese, "Health Unit Warns Against Using Trout Lake Water Near Chemical Spill" *The Nugget*, May 21, 2012. http://www.nugget.ca/2012/05/21/health-unit-warns-against-using-trout-lake-water-near-chemical-spill. Also P.J. Wilson, "Crews Neutralizing Formaldehyde," *The Nugget*, November 5, 2012. http://www.nugget.ca/2012/11/05/crews-neutralizing-formaldehyde

36 "We [The City of North Bay] have applied to the National Energy Board for intervener status. (The National Energy Board will be holding hearings on the pipeline project). Our yearly operating and capital budget here in North Bay is about $150 million. We've set aside $250,000 for environmental and legal expertise, which is not a lot of money when you make an application to the NEB. We weren't going to stand by and let others speak for us. We want to be at the table to express our concerns. Having the Ontario Energy Board come out with this report gives us comfort and confidence." Mayor McDonald as quoted in Donovan Vincent, "North Bay Mayor Echoes Concerns About Energy East," *Toronto Star*, August 14, 2015. https://www.thestar.com/news/canada/2015/08/14/north-bay-mayor-echoes-concerns-about-energy-east.html.

37 Kelly Anne Smith, "Nipissing First Nation Set to Develop Position on TransCanada's Energy East Pipeline," *Anshinabek News*, March 23, 2016. http://anishinabeknews.ca/2016/03/23/nipissing-first-nation-set-to-develop-position-on-transcanadas-energy-east-pipeline/.

38 The cabaret included musical and dance performances, video shorts, and poetry by artists including the Hidden Roots Collective, Aanmitaagzii, Peter Cliché, Gary & Rene White and Rob Joanisse, David DinoWhite, and the Blackbirds. Stop Energy East (North Bay). http://www.stopenergyeastpipeline.ca/.

39 This rally was primarily concerned about the deregulation of drinking water through federal legislation, which would mean less protection for Trout Lake with the proposed pipeline conversion. Maria Calabrese, "NDP Rallies for Trout Lake," *The Nugget*, July 31, 2013. http://www.nugget.ca/2013/07/30/ndp-rallies-for-trout-lake.

40 See for example, Sam Buzzo, *Blue Gold: World Water Wars* [DVD], Maude Barlow and Tony Clarke actors (Mongrel Media, 2009).

41 On "glocal," see for example, Paul Virilio, "Speed and Information: Cyberspace Alarm!" *CTheory*, August 27, 1995. http://www.ctheory.net/articles.aspx?id=72.

42 Thomas Andrew Gustafson. 2013-10-28. "How much water actually goes into making a bottle of water?" *National Public Radio [NPR]*, October 28, 2013. http://www.npr.org/sections/thesalt/2013/10/28/241419373/how-much-water-actually-goes-into-making-a-bottle-of-water.

43 Ice Follies website: http://icefollies.ca/home/2016-theme/.

44 Aanmitaagzi website: http://aanmitaagzi.net/. Also Near North Mobile Media Lab: http://www.n2m2l.ca/ and White Water Gallery: http://whitewatergallery.com/.

45 Edgardo Moreno's *Nisidotan* was presented by Zakide, an Aboriginal artist-run center operating out of various locations in Northern Ontario (Canada).

46 *Serpent People* was a workshop performance hosted by Aanmitaagzi, a community-engaged artist-run center (Nipissing nation).

47 Jihee Min's *A Humble Trowling* was presented by Gallery 44 (Toronto).

48 Meg Paulin's *The Trapper's Cabin* was presented by Aanmitaagzi.

49 Darren Copeland, *Fishing for Sounds*, presented by White Water Gallery (North Bay).

50 *STAG* (*Student Temporary Art Gallery*) was curated by Imogen Wilson and presented by Nipissing University.

51 Caitlind r.c. Brown & Wayne Garrett's *The Deep Dark: Crossing* was presented by N2M2L, the Near North Mobile Media Lab (North Bay).

7 Facing the Elemental
Amanda Boetzkes, Emmanuel Levinas and *Shooting Theory*[1]

Shannon Bell

"Facing the Elemental" arises out of the seventh installment in an on-going project *Shooting Theory*[2] which brings digital video technology and print textual theory together through imaging philosophical/theoretical concepts. The overarching argument is that one ought not to think political theory simply within language.

Amanda Boetzkes's work on the elemental is a significant *ex post facto*, after the fact, influence on my image-text "Facing the Elemental" presented here and my film *Shooting the Elemental* from which this image-text is derived.[3] The idea of "ex post facto influence" is one way to imagine how theory can connect thinkers whose work shares resonance, even while sometimes taking dramatically different forms.

Boetzkes establishes a connection between earth art and the concept of the elemental as this is elaborated by Emmanuel Levinas, John Sallis and Alphonso Lingis. For Boetzkes, the overarching question concerning the elemental is: "[Does] the elemental only surface other beings and other things? Can the elemental yield the earth as a face, as an unnamable alterity, and as an imperative? Can we receive the elemental as a means of giving the earth a voice without anthropomorphizing it[?]"[4] Boetzkes's answer is that artists do precisely this in creating works that are bound up in the elemental but that also yield sensation through gestures that simultaneously offer the body and infer the withdrawal of the subject's perceptual expectations."[5] It is with this trajectory in mind that I visualized the elemental in my work with the Alaska landscape, in which I did site specific theory, conceived as a form of linking ideas directly to the elemental qualities of the land.[6]

Facing the Elemental/Shooting the Elemental was shot/filmed from the air, water, and earth of Alaska, specifically around Aleyska Highway, Girdwood, Whittier and Prince William Sound. Levinas reflects on how air, water, wind, sky and sea anonymously compose our enjoyment of "something." We dwell in the elemental in our daily living grounded on earth; our "bathing in the element," in which the finite human component is non-distinct from the infinite, is most often taken-for-granted and not evident.

Levinas's concept of attention combined with Simone Weil's concept of attention were relayed to the visage or face of mountains and glaciers from the elemental sites of air and sea. The aim is similar to that achieved by Alphonso Lingis, in *The Community of Those Who Have Nothing in Common*, where he shot the face of the other from "the point of departure" of the face being photographed.[7] Only in my case the visage is the elemental manifest as the side of mountains and glaciers, "the edge of the wind" on the surface of snow and on "the surface of the sea."[89] The objective is to face the natural objects of mountains, glaciers, waters with the consciousness that I am a "condensation of earth, light, air and warmth."[10]

The aim of the endeavor (work and attention) is to visually present that which cannot be captured except via the trace which it leaves behind on the edge, surface, side. The film is shot in color; the shades of grey, white and in-between are the colors I was seeing as an existent moving through existence. As Levinasians know the elemental can only be seen as a fleeting side, surface, edge always threatening to recede into the *il y a*, or the anonymous existence of the "there is."

This is a brief descriptive review of Levinas' elemental developed through reading his book *Totality and Infinity* and John Sallis's "Levinas and the Elemental"[11]:

1. The elemental is the background medium [milieu] in which everything takes shape including ourselves: "one is steeped in it; I am always within the element."[12] One exists "bathing in the elemental."[13] Or as the American poet Robinson Jeffers puts it: "I wander in the air, Being mostly gas and water, and flow in the ocean."[14]

2. The medium is non-possessable, "nobody's": earth, sea, light[.][15]

3. "The medium has its own density."[16] It is depth and content without form.

4. When one comes face to face with the elemental it is always as a side, surface, edge; that is, we touch the face of the elemental when we encounter "the surface of the sea and of the field, the edge of the wind."[17]

5. The elemental refuses representation—it cannot be fixed as an object. "It is wind, earth, sea, sky, air … it precedes the distinction between the finite and the infinite."[18] Levinas refers to the elemental as "an opaque density without origin."[19]

6. The elemental is our domicile; and what gives us a foothold in the elemental is labor. Levinas cites the action of cultivating a field, fishing, cutting wood.[20] He also indicates that "the aesthetic orientation that humans give to the whole of their world represents a return to enjoyment and to the elemental on a higher plane."[21]

7. The elemental can never be revealed beyond its visage—side, surface, edge—because what the side–surface–edge conceals "is not a 'something' susceptible of being revealed, but an ever-new depth of absence, an existence without existent, the impersonal par excellence."[22]

8. The depth of the element prolongs the [visage from multifold sites] till it is lost in the earth and in the heavens. "Nothing ends, nothing begins."[23]

Like Husserl's *epoche* and Lacan's Real, imaging the elemental is seductive and elusive. It seems to me that the only ethical way of approaching the elemental is through attention. Attention to both the objects that present themselves to our consciousness and to consciousness itself is the core of the phenomenological method. My use of Levinas's concept of attention (infused by thoughts from Simone Weil) aligns with the idea of phenomenological intentionality that Boetzkes describes as the moment that the intentionality of the subject's attention is opened to disruption by its object. With reference to James Turrell's *Roden Crater*, Boetzkes observes:

> Turrell's practice problematizes the commonly held view that the aim of earthworks is to elicit active and intentional gestures of seeing and interpreting, for his installations and light chambers often require one to stand still, sit, or lie down in a particular

> place in order to witness a play of light, sound, or atmosphere. This is not to say that the viewer is no longer an active, intentional, embodied being; rather, the artwork disrupts that movement and intention in order to fully open the senses to the elemental ....[24]

What Levinas and Weil add to the Husserlian concept of attention is an understanding of attention not only as the process of consciousness but also, as Boetzkes argues, attention as an ethical phenomenon, "a reciprocal relation between oneself and the object or other person in the visual field."[25]

Michael Marder, "A Levinasian Ethics of Attention," contends that Levinas "assigns to attention a crucial role coextensive with Husserl's intentionality; Levinas recognizes in attention a "subjective modification" of intentionality: "being attentive [to] the call of the other."[26] Marder maintains that the "Husserlian reduction is not radical enough for Levinas's philosophical taste, since it fails to recognize that this life comes into being thanks to the appeal emanating from the Other, whose calling out to me forces me to pay attention ...."[27] Attention for Levinas, read through Marder, is "the condition of possibility for thought, consciousness itself and what he terms 'the reduction to the ethical'."[28] Attention to the other is "the point of departure"; "Attention is attention to something because it is attention to someone."[29]

Levinas does not, however, set forth any method of engaging in such attention. It is here that Simone Weil's concept of attention necessarily supplements Levinas'.

For Weil there are five conditions that define the process of attention:

1. "Attention, taken to its highest degree, is the same thing as prayer .... Absolutely unmixed attention is prayer."[30]

2. "Extreme attention is what constitutes the creative faculty in [hu]man[s]."[31]

3. "[A]ll that I call 'I' has to be passive. Attention alone—that attention which is so full that the 'I' disappears—is required of me. I have to deprive all that I call 'I' of the light of my attention and

turn it on to that which cannot be conceived."[32] "Attention … is so full that the 'I' disappears."[33] Or, as Levinas says: "in attention the I transcends itself."[34]

4. "Attention consists of suspending our thought, leaving it detached, empty and ready to be penetrated by the object[.]"[35]

5. "Creative attention means really giving our attention to what does not exist."[36]

The process of attention (focusing on the exteriority of what does not exist, emptying thoughts, bracketing or making the "I" disappear) is turned to depth that cannot

be approached except in specific notations of the visage—side–surface–edge—of rock, water, air. The process of attention is a labor of attention that gives us access to the "opaque density without origin" as a trace on the—side–surface–edge—until again "it is lost in the earth and in the heavens,"[37] just "as water gushing forth from rock washes away that rock."[38]

Of course, work, for Weil, is time entering into the body and for Levinas it is the action that grounds us in our domicile, "the way of access to the fathomless obscurity of matter."[39]

Extreme attention focused on the other that cannot be conceived, that can only be come up against, thoughts emptied, the "I" bracketed, camera pointed: the meditative count is 10 breaths, shoot for 30 seconds, stop; look, wait or begin immediately again, 10 breaths, shoot for 30 seconds—done 10 times in the air, 10 times on the water and 10 times on the ground for a total of 900 seconds or 15 minutes.

In *Facing the Elemental*, the body, my body, acts as a conduit *with* the elemental, a manifest part, which reveals surface visages from its location/home in the elemental. While these visages can be presented, what resists representation, having presence only as a trace, is *il y a,* the anonymous existence of the there is, which we—body, air, water, wind, sky, breath—belong to.

Facing the Elemental offers, in Boetzkes terms, "a reciprocal touch between the body and the earth" that attempts "a way of seeing the earth by receiving its 'gaze',"[40] being a part of what is looked and looking back through the act of "facing the elemental" with attention.

Image Credits

Shannon Bell, image stills from: *Shooting the Elemental*, 2012. Video. https://vimeo.com/42109865

Notes

1 I would like to thank Raan Matalon for his research on this project and for the time he has given to reading and discussing the elemental and attention with me. As always, I wish to thank Gad Horowitz for his theoretical and editorial assistance and for his bringing to my attention the work of the poet Robinson Jeffers and the photographer

Morley Baer who consistently come face-to-visage with the elemental in their words and images.

2 Thus far, I have produced twelve film projects as conceptual theory:

- Martin Heidegger—*Dynamic Stillness*, 2007. 40 days and nights of sunrise and sunset, Judean Desert;
- Edmund Husserl—*Epoché Reflections* and *Blind Residuum Caves*, 2008. Judean Desert, Negev, Golan, and Galilee. https://vimeo.com/21341729;
- George Bataille and Simone Weil—*Beautiful Waste: Dead Sea Sinkholes,* 2009. Dead Sea. http://www.youtube.com/watch?v=CDtr_8n54Cg http://www.youtube.com/watch?v=MrFfwa483A4;
- Paul Virilio—*Camel Vision Machine*, 2009. Wadi Ram: Jordian Desert;
- Gilles Deleuze—*Shooting the Blur*, 2010. The Syrian state, the Israeli state, the British state, the Jordanian state, Golan, North shore of the Dead Sea, Abu Dis, Jerusalem, Gilo/ Beit Jala. https://vimeo.com/24035994;
- Jacques Lacan and Paul Virilio—*Imaging Time and the Real*, 2011;
- Emmanuel Levinas—*Shooting the Elemental,* 2012. Aleyska and Seward Highways, Girdwood, Whittier and Prince William Sound\ Lake George Valley, Lake George Glacier, Colony Glacier, Whiteout Glacier and Surprise Glacier, Alaska. https://vimeo.com/42109865;
- Walter Benjamin—*Flâneuring Ancient Arcade Ruins,* 2012. Israel/Palestine. https://vimeo.com/46174007;
- Martin Heidegger—*Flashes of Perception*, 2013. Pangnirtung, Cumberland Sound, Saniru in Baffin Island, Eastern Arctic. https://vimeo.com/71183979;
- Samuel Mallin – *The Sinuous Turn*, 2013. Ancient Minoan lines, Gournia, Knosos, Malia, Phaestos in Crete. https://vimeo.com/77701915;
- Alain Badiou and Hindu Theory of Numbers—*Shooting Zero in ancient lands,* 2014. Salt Icon of Zero in Israel and India. https://vimeo.com/115414403;

- Georges Bataille—*Expenditure in Mount Etna*, 2015. https://vimeo.com/150547004.

3 *Shooting the Elemental*, Shannon Bell, 2012. https://vimeo.com/42109865

4 Amanda Boetzkes, *The Ethics of Earth Art* (Minneapolis: University of Minnesota Press, 2010), 161.

5 Ibid.

6 I did this film project for the seventh annual North American Levinas Society (NALS) conference in Anchorage, Alaska; the conference theme was the elemental.

7 Emmanuel Levinas, *Totality and Infinity. An Essay on Exteriority*, trans. Alphonso Lingis (Pittsburgh University Press: Duquesne University Press, 1969), 99.

8 Ibid., 131.

9 Shooting in HD the conceptual image will pickup both the "background noise of existence," as Lingis is fond of saying, and beyond what the human eye can see. Alphonso Lingis, *The Community of Those Who Have Nothing In Common* (Bloomington and Indianapolis: Indiana University Press, 1994), 43.

10 Lingis, 122.

11 See John Sallis, "Levinas and the Elemental," *Research in Phenomenology*, N. 28 (1998), 152-159.

12 Levinas, 131.

13 Levinas, 132.

14 Robinson Jeffers, *The Collected Poetry of Robinson Jeffers: 1928-1938,* ed. Tim Hunt (Stanford California: Stanford University Press, 1988), 124.

15 Levinas, 131.

16 Ibid.

17 Ibid.

18 Ibid., 132.

19 Ibid., 158.

20 Ibid., 132.

21 Ibid., 140.

22 Ibid., 142.

23 Ibid., 131.

24 Boetzkes, *The Ethics of Earth Art*, 111.
25 Ibid., 28.
26 Levinas, 178.
27 Michael Marder, "A Levinasian Ethics of Attention," *Phainomenon*: Revista de Fenomenologia, 18/19, Spring & Fall 2011, 27.
28 Ibid., 30.
29 Levinas, 99.
30 Simone Weil, *Gravity and Grace*, trans. Emma Crawford and Mario von der Ruhr (London and New York: Routledge, 2002), 117.
31 Ibid.
32 Ibid., 118.
33 Simone Weil, *Notebooks of Simone Weil*, vol. *1, trans.* Arthur Willis, (London and New York: Routledge, 2004), 179.
34 Levinas, 138.
35 Simone Weil, *Waiting for God*, trans. Emma Craufurd (New York: Harper Collins Perennial Classics, 2001), 62.
36 Ibid., 92.
37 Levinas, 131.
38 Ibid., 127.
39 Ibid., 159.
40 Boetzkes, *The Ethics of Earth Art*, 160.

8 An( )Alibic Aural Tetrad
A Fourfold Structure of Ecologicity

David Cecchetto

I—Distributed Allures I (Having Doubly Been Where One Wasn't)—Dampen!

There is a propensity for pedestrians speaking on mobile phones to be struck by vehicles not because they walk into traffic *per se*, but because they escape the driver's visual field by talking on the phone. That is, this class of accident happens not because the pedestrian walks into traffic but rather because the pedestrian fails to share the customary, nonconscious affective cues of recognition with the driver such that their particular vitality in turn disappears into the driver's traffic environment. The communicative tool works perfectly, and the pedestrian disappears into its working.

We might take this disappearance not just on its face, but also as evidence of a certain situational priority of the mobile phone over the pedestrian; as evidence of the extent to which the situation is one in which the pedestrian is existentially absorbed into their communicative tool. That is, the pedestrian-mobile phone coupling is a self-contained system without a discernible break. This is evident because if there were a visible break—if the pedestrian's humanity, for example, were radically incommensurable with this coupling—then there would also be a disclosure of the pedestrian as pedestrian. Presumably, such an existential swerve would beget a more literal one enacted by the oncoming car, with disaster avoided. Instead, the perfect functioning of the coupling as communicative tool conceals the "*as*-structure" of the human-mobile phone coupling, which is to say the coupling *as* coupling (i.e., as a conjoining of incommensurable object existences beneath the smooth surface of their functioning).[1] The tool works by absorbing the pedestrian as pedestrian into its workings. In this sense, the pedestrian's death (by traffic accident) is caused by their not being present at the scene of their death; if they'd been there, they wouldn't have been killed (or, at least, they would have been killed differently).

The pedestrian's disappearance, though, is not really a single movement into a tool, but rather the revelation of a distributive ecology composed of "overlapping material conditions" that can't be collapsed into a "located place or closed system."[2] It's true: the pedestrian *goes away,* disappearing into the mobile phone's global systematicity. However, it is also the case that the pedestrian *arrives* at their tele-location(s), at the post-global technical ecology of cellular, satellite, and smartphone technologies that both exists in its own right and textures contemporary sociality. The pedestrian appears, then, but differently and distributedly in the most radical senses. One person's disappearance is that same person's appearances (as the saying should really go), which is to say that appearance and disappearance are neither opposites nor symmetrical, but something closer to locations; or rather—since they are multi-vectoral and radically contingent—locationalities. Put differently, (dis)appearance discloses a performative—which is to say radically excessive—circuit through which incommensurable objects relate even as they remain discrete, withdrawn in an openness that displays their sum to be less than their parts (but more so).

Such a circuit conforms to Graham Harman's "total empire of equipment," which describes the world as a "single system of tool-being" that "absorbs every being" such that "there can never be any singular end-point within the contexture of reference."[3] To consider the ill-fated pedestrian, though, is to reverse perspective on this in order to emphasize the relational, sensorial dynamism through which the problem of a singular end-point becomes alluringly thinkable in the first place. That is, the disjuncture that characterizes the pedestrian's locationality isn't opposed to a single system, but is rather the phenomenal outcome of it. The pedestrian is continually relocated, but "this occurs through the relocation of the interaction between the subject and the environment to a surface that elicits the retraction of the subject and is receptive to a manifestation of [phenomena that are withheld from being known but also have sensorial plenitude]."[4] Put differently, the pedestrian's relocation takes place by virtue of being multiply located and thus subject to collapse into any given location/system by that system's operations, and this "subject to" discloses the ontological priority of the multi-locationality by "direct[ing] perception toward the sensation of [the total empire of equipment]."[5]

And indeed, the circuiting that is, comes from, and becomes the sensorialism of the pedestrian-mobile phone coupling works precisely to demonstrate the importance of understanding not just the way things really are (i.e., ontology), but equally the "way things really aren't."[6] And really, this anticipates the final section of this short essay, which considers the doubling causalities disclosed in "high-place phenomenon" as a sensorial description-prescription: to the extent that paradoxicalities of agency must be met speculatively, they in fact *demand* to be approached supralogically, which is to say through 'pataphysical techniques: the lures of what Daumal calls "prospecting" and the attentional force of his 'pataphysical drilling are necessary in order to hear this appearance-ness in its audible profile.[7] One can hear metaphysics 'pataphysically, but only because its metaphysical ringing is dampened in advance. What, then, is heard in and as the world's dampening? What is heard as the dampening of the world? And what are the resonant frequencies of the differences between those dampenings?

II—N.B.—Listen!

It almost goes without saying: to listen is to acknowledge the world in its "ecologicity," to call the world forth as a qualitatively singular constellation of objective conditions and mobile sensual effects.[8] And yet, in so far as listening involves attention it is equally (though not more) about misdirections—always more than one at a time—as it is about any conventional understanding of focus; that is, it is about the material misdirections that are the calling forth as the performative excesses of constellating, objectifying, conditioning, mobilizing, sensing, and effecting. *NB: Materiality is always in performance, and performance is always productive of excesses.*

In their own ways, musicians will tell you as much, repeating—for example—Debussy's dictum that music is found in the spaces between the notes. Indeed, the challenge of playing in an ensemble might be characterized in this way too: one must listen simultaneously to oneself and the ensemble in both their collectivity and their distinctness, the former for obvious reasons and the latter because one must nonetheless play one's part with the specificity that both is and signals "musicality." Sing it in a round: musicality as circular causality. *NB 1: A round isn't actually circular, it's one of those cases where something is cited as relatively more complex than it might be—e.g., a round conjures musical time as a spiral rather than a line—and in so doing foreclose on its more radical complexities (e.g., that music may not be spatial at all); NB 2: Circular causality isn't actually circular, which is why one ends up thinking about listening in terms of ecologicity (i.e., in order to account for the radically singular* quality *of circular causality that is foreclosed by the relativity packed into the adjectival use of "circular").*

Listening is thus a recursive process of (non)selection and thriving. Even in a more limited field, though, such listening—which is all listening, not just musical listening—is not *about* selection, per se, in that one's (for example) listening away from oneself to a collective isn't in opposition to listening to oneself. Rather, listening is listening in so far as when one listens one attends to that of a sound which is not sounded, which is to say one listens to music in its nonlinearity (i.e., as a system that outputs signals that are qualitatively different from its inputs). One listens *to and away:* the sum of all possible attendances is less than its parts, but that less is precisely also (and more importantly) *more* in that its resonant affordances continually reinforce themselves. Sounds have much to say, but they don't say it . . . they say something else. Put differently, the sum of all the musical sounds present in a room is less than its parts, but more so. *NB: Reality is a room, among other things; a room is also a room, among other things (as Inspector Clouseau's requests for one reveal).*[9]

Listening, then, is (in)attention. Importantly, though, this (in)attentional economy in no sense operates in the sole or even privileged mode of conscious thought. The (in)attention of listening is, for example, played out in and as the physiology of the ear itself: on one hand, it is simple enough to understand the transition of sound energy from the relatively large—and, indeed, airy—outer ear to the tiny oval window that acts as a threshold to the fluid-filled inner ear as precisely an attentive process. That is, the middle ear functions primarily to concentrate—to focus—the pressure exerted by a sound wave onto an eardrum into an area (i.e., the oval window) that is approximately twenty times smaller than it, thus working rather like a thumbtack. On the other hand, though, the mechanical coupling through which this takes place is rather more complex because it occurs via *not one but three* causally successive bones, the interaction of which allows for—or, put less psycho-centrically, causes—various regulatory functions. Thus, as one example of many, when the middle ear's stapedius muscle contracts it reduces the motion of one of the three bones (the stapes) in a manner that affects the transfer of some frequencies more than others.[10] *NB: "Transfer" is a term of (in)convenience, purposely chosen over "transduction" because the latter, in being slightly more accurate, might seduce one into forgetting that the entire causal chain—in being called forth as such—occludes the radical relationality that is in play; that is, occludes the weird, alluring temporalities of ecologicity to which listening listens.*

We listen in part by not listening. Listening is "the contraction of all sound, the contraction of all vibrations, which gives sense to sound, contracting clearly just *this* vibration, *this* sound wave, and letting the rest remain obscure, implicated in various degrees of relaxation."[11] And, while one might think—in concert with an informatic logic that imagines communication to consist in point-to-point transmissions of data—of this as a simple filtering process, the physiological fact of the matter is that one relies on the *dynamism* of the middle ear as much as its filtering profile. Put differently, since one only hears via the contractively transductive process of hearing, and since that process is inseparable from the specific and material misdirections of the middle ear's dynamism (among other dynamisms), it follows that to listen is to attend to the effects of a reality the cause of which can never be singly determined, even as a coming together of more than one. *NB: The proverbial sound of one hand clapping is not the limit case of sound, but rather its basic enabling condition … providing that we accept that every hand is singular precisely because it is itself a multiplicity.*

Put differently, the ecology called forth in listening always includes an autonomic, oto-acoustic dimension; specifically, it always includes the ongoing and relentless dynamism of intra-ear relations. Thus, while it is true that one breaks a physical transmission in order to have received it, it is more importantly the case that one (materially) conceives a transmission such that one can hear the ongoing relations (the contraction and dilation of the stapedius, in concert with innumerable other processes, the separation of which—i.e., the framing of such processes as distinct processes—is always contingent) … or rather, in order to take part in the transductive energetic constellation that allows for questions of meaning(lessness). The ear functions in communication in the form of an alibi, dissimulating its ecologicity in order to function, with the particularity of any given instance of "functioning" acting to "disclose [determinable] signals of an otherwise [undeterminable] object world."[12] Indeed, this is precisely why it is so important to listen well, as this alibic function is as much evidence of listening's (and, indeed, any) communicative importance as one is apt to hear. *NB: Tinnitus is also not an exceptional case with respect to listening, but rather a basic enabling condition. One listens tinnitally to the clapping—the successive impulses—of a singular multiplicity. Listening thus signals sound's migration beyond its enabling conditions, namely changes in air pressure.*

Like I said, this almost goes without saying. Sometimes, though, saying something can work to bring forth what is said as a thing in its own right, which is to say as a before *and after* of its objective material existence.[13] What then, is the thingness of listening? If listening is constitutively misdirected—if it is a radically contingent production—then such a question can only be answered according to specific instances, otherwise the misdirection would be relativized. Moreover, to listen to listening would require a misdirection in its own right, a second-order of misdirection; it would require one to listen to one's listening, the ensemble of listenings, and their summing that is less than their parts (but more so).

III—Quasi-causal Superhighways—Speak!

In "Interpretation and the Affordance of Things," Amanda Boetzkes provocatively reads Nam June Paik's *Electronic Superhighway, Continental U.S., Alaska, Hawaii* as revealing of a certain historicity of video. This grand-scale 1995 installation is:

> composed of 313 monitors of all different sizes, projecting 51 channels, set into a neon-tube map

> of the United States. Stacked together in clusters that approximate the shape of each state, the televisions are shelved awkwardly, some resting precariously on their sides, others shoved into small spaces, bearing the weight of much larger monitors. All are wired, and together transmit a panoply of images and sounds. The title of the piece is in part a literalization of the electronic age as "an information superhighway," a term which Paik claimed to have coined in 1974, long before the phrase became a popular way of describing the possibilities of the Internet. The work is a play on McLuhan's reflection that in the electronic age, light is a primary medium, which is to say that electricity produces a specific distribution of sense that is constitutive of our patterns of communication. Thus, electricity was the condition of possibility for an information network parallel to a superhighway.[14]

The reading—the interpretation—is convincing and compelling: by tracking *Information Superhighway* in this way Boetzkes reveals its revealing, namely the way that the indelible objects in *Electronic Superhighway* not only "disclose [the] equipmental basis of the medium but also underscor[e] that technology is perpetually advancing and redeploying itself, and in so doing, prescribing a history according to its own development."[15] This aligns, for Boetzkes, with the Heideggerian description of the historical movement of technology as *destining.*

Put differently, Boetzkes reveals through Paik a historicity of video (and television) that exceeds the technology per se. As she notes, "despite the inextricability of art and technology, art is never wholly concealed by technology, but instead juts out as an obtrusive thing, like an iceberg from the flow of history."[16] Thus, Boetzkes continues her argument—her interpretation—by considering another work by Paik, *Family of Robots* (1986), which is comprised of:

> anthropomorphic sculptures made from television boxes, each becoming a member of a family lineage of three generations. The grandparents on one end are made out of vintage television and radio casings, their faces suggested by the knobs on transistor radios. Screen surfaces, varieties of rabbit ears and the edges of the casings demarcate limbs, joints and body parts. On the other end of the evolution, the *Hi-Tech Baby* is made out of 13 aluminium television casings, a more modern-looking material. All the figures are wired to a single channel, and remain silent.[17]

For Boetzkes, Paik's foregrounding of the television casing and attenuation of the medium's function "calls attention to affordances that have always been there, but which remain unseen if the television is broadcasting (that is, when it is absorbed into the machinations of equipment)."[18] Somewhat counterintuitively, then, the historicity of the technology is not only revealed in its nonfunctioning, but specifically in a *history* of its extra-functionality.[19] We are given, then, a succession of events (the history of these technologies) that are causally connected without being determined, wherein the indetermination of the connection lies (paradoxically) precisely at the site of the "progressive," causal impingement that one technology exerts on the next. Put differently, the historicity of this technology describes a history that unfolds via a causal logic that is at once distributed and reversible, which is to say a logic that is not properly causal at all but rather quasi-causal.[20]

Notably in this context, Boetzkes's interpretive argument doesn't just describe a bunch of things, or even thingliness, but is also itself a *thing:*

> [It] propels thinking beyond the parameters of any causal relationship between object and meaning, or instrumental relationship between form and function, and instead reveals objects in their operation of mirroring, ringing, and conjoining a world and our sense of it.[21]

And it is this thingliness of the argument—this interpretive gambit through which Boetzkes speaks knowingly of precisely that which cannot be known—that affords the emergence of the historicities that it describes. That is, the artworks, technologies, thoughts, and creative gestures in question are historical, but their historicities only become palpable through the ontological openness—the constitutive porousness of an object that signals its withdrawn essence—of their dynamic, collective sensibility.

Thus, Boetzkes's interpretative gestures are precisely more than the final interpretation itself, sounding something deeper: the translative profile of allure that guarantees the latter's bypassing of any human/nonhuman distinction. That is, while the interpretation itself issues from Boetzkes, the interpreting framework recursively constitutes her as an occasion of thought's constituent operation such that she "is the final end whereby there is the thought" to exactly the extent that the thought is hers.[22] In short, the affordance of the situation emerges through a reiterative misdirecting of a possessive inversion. One can hear it, if one listens. (It is entirely possible that "one" can never listen—i.e., if "*the relation* is the smallest possible unit of analysis"[23]—but consideration of that is for another moment; I mention it here only to remain with the 'pataphysics of the problem at hand.)

So, a question: if one (misdirectedly) folds Boetzkes's argument into the art-theory coupling she describes, what does one hear? One vibrational frequency that becomes audible is the specific way that the "first-ness" of Graham Harman's "aesthetics as first philosophy" takes the form of alibi when worked through on this terrain. That is, the readings of Paik (and others) deepen the broader engagement of Boetzkes's piece, which teases out and weighs "the notion of affordance against disciplinary assumptions about intentionality, meaning, and context" via provocatively mashing up the artworks she examines with Harman's Heideggerian tool-analysis.[24] In so doing, Boetzkes's essay neither undermines nor overmines (to use Harman's terms), which is to say that she neither uses *Electronic Superhighway* to exemplify the topology she is excavating nor works "backwards" from an unmediated phenomenological encounter with the artwork to build out her theoretical position. Instead, Boetzkes *gathers* the artworks and her theoretical levers. And this gathering signals a meeting point of metaphysics (i.e., "first philosophy") and aesthetics in so far as both have to do—though differently—with the way things "hang together" (in the parlance of Whitehead). And yet, it is manifest that this gathering takes place—that it resonates in a thingliness that at once occasions its appearance and its fundamental openness—as a singularity that signals the operation of aesthetics, even as it dissimulates that singularity by being entirely reasonable. What, then, is the terrain of this gathering? The (in)sensible field through which figurations figure in their fully textured specificity?

IV—Distributed Allures II (Having Not Been Where One Doubly Was)—Resonate!

In a 2012 study, the psychologist Jennifer Hames and her collaborators coined the term "high-place phenomenon" (HPP) to describe the apparently quite common urge to, for example, jump off the edge of a proximate cliff despite not really wanting to kill oneself.[25] In their study, about a third of the sample said they'd felt the urge to jump from a cliff—or drive into oncoming traffic or something similarly suddenly lethal—at least once in their life, and while clinically depressed people were more likely to feel this, over 50% of those who experienced the phenomenon said that they'd otherwise never considered suicide.

When the results of the HPP study were correlated, Hames and her team of psychologists speculated that the suicidal urge that so many people experience never quite happens as such. Instead, they suggest a scenario wherein a sensitive and anxious person—and HPP tends to correlate with anxiety—leans over, say, a ledge and doing so causes a super-fast instinctual reaction to the physical sensation of anxiety such that the person's survival instinct forces them to move away from the edge. When this person looks at the ledge from which they've just recoiled, though, they see that it is sturdy and that, as such, there was never any real danger. Thus, they are met with the question, posed by a mind that insists on linearity, of why they backed up if the ledge never posed a threat … and the suicidal allure is produced, *post factum*, as the most plausible, logical answer to that question. In other words, Hames's hypothesis is that people misinterpret an instinctual *safety* signal and thereby conclude they must have felt an urge to leap. They leap into the past to produce a cause to the manifest effects of a lived anxiety. It's suicide as time travel in the form of an imagined death that is less than the sum of its parts, but more so.

In Hames's view, it's easier to think that one thought something one didn't than it is to think that there might be something unthinkable in play. Or further, presuming one might have an experience of HPP subsequent to knowing how it works: it's more difficult *not* to think that one thought something one didn't—more difficult to give oneself fully to the non-thinking of thought—than it is to imagine that the network that connects thought, instinct, consciousness and action might not fully schematize … an imagination that is itself seemingly impossible at the level of instinct. Time is always temporalized in material processes—it always abides by what Man Res Khan calls the "law of temporal finitude"[26]—and in this case the time of the suicidal allure materializes as the non-coincidence of instinct and thought: the urge to jump is an apocryphal smoothing out of the irregularities through which a decision will have been made.

For Hames, this means that the perceived urge to jump in fact *affirms* the urge to live, which is indeed the title of the study. But something like the reverse is also true, then: if Hames's hypothesis holds, the a-rationality of instinct—the withdrawal from a danger that was factually not there—suggests an instinctual allure to the self-destructive affordances of a ledge; an instinctual awareness of the way that agencies are couplings of humans and nonhumans alike, and that ledges and gravities have instincts of their own to which they attend.

Listen: agencies are distributed and recombinant. Listen: actions are not causes of effects, but are rather palpations that *resonate* along different and varying catalytic vectors. Listen: the question lies in both how one attunes *and* how one learns to attune to this multiplicity. Listen: how to listen?

Notes

1 Amanda Boetzkes, "Interpretation and the Affordance of Things," in *Heidegger and the Work of Art History*, eds. Amanda Boetzkes and Aron Vinegar (Burlington, VT: Ashgate Publishing, Ltd., 2014), 273.

2 Amanda Boetzkes, *The Ethics of Earth Art* (Minneapolis: University of Minnesota Press, 2010), 143.

3 Graham Harman, cited in Amanda Boetzkes, "Resource Systems, the Paradigm of Zero-Waste, and the Desire for Sustenance." *Postmodern Culture* (forthcoming: 2017).

4 Boetzkes, *The Ethics of Eart Art*, 142. For the sake of clarity I have substituted the bracketed text for the original "elemental phenomena." While I think this substitution is in the spirit of the original (because it works from descriptions of the elemental in the cited text) it also introduces a certain slippage that, while productive for my argument here, redirects something of Boetzkes's argument in the book as a whole in ways that are not altogether adequate. The same is true of the bracketed text in the following sentence.

5 Ibid. See previous note for explanatory comments.

6 René Daumal, *Pataphysical Essays*, trans, Thomas Vosteen (Cambridge, Mass: Wakefield Press, 2012).
7 Ibid., 28.
8 Boetzkes, "Interpretation and the Affordance of Things," 275.
9 https://www.youtube.com/watch?v=OsCW2N-6MUM
10 This contraction most often occurs as an unconscious reflex when one is exposed to loud sounds, thus protecting—though often belatedly, because it is slower than the speed of sound—the relatively delicate structures of the inner ear.
11 Aden Evens, *Sound Ideas: Music, Machines, and Experience*, Theory out of Bounds, v. 27 (Minneapolis: University of Minnesota Press, 2005), 16.
12 Boetzkes, "Interpretation and the Affordance of Things," 288. I have substituted "determinable" and "indeterminable" for Boetzkes's use of "visible" and "invisible" in order to avoid certain confusions. While this substitution aligns—to my mind—with her argument in this case, this is not to suggest that it obtains more broadly. Clearly, Boetzkes's work—in the cited chapter and elsewhere—works through the operations of (in)visibility in concert with specific aesthetic regimes of visibility as well as specific ocular and neuroscientific discourses related to the eye, none of which nuance would be captured in the terminological substitution I've made here.
13 Ibid., 271 (emphasis added).
14 Ibid., 276.
15 Ibid., 277.
16 Ibid., 280.
17 Ibid., 278.
18 Ibid., 280.
19 The undecidability of the possessive here is revealing of a certain fault-line: can extra-functionality ever belong to that to which it is extra? And yet, how can it be thought without such a belonging?
20 Perhaps "post-causal" is more to the point, since this is causality's immanent critique of itself read through its

vectors of productivity. Alas, that prefix is a step too cringe worthy to adopt … aesthetics as first nomenclature!

21 Boetzkes, "Interpretation and the Affordance of Things," 271.

22 Alfred North Whitehead, *Process and Reality*. 2nd Revised edition. (New York: Free Press, 1979), 151.

23 Donna Haraway, *The Companion Species Manifesto: Dogs, People, and Significant Otherness*, (Chicago: Prickly Paradigm Press, 2003), 20 (emphasis added).

24 Boetzkes, "Interpretation and the Affordance of Things," 270.

25 Jennifer Hames, Jessica D. Ribeiro, April R. Smith, and Thomas E. Joiner, "An Urge to Jump Affirms the Urge to Live: An Empirical Examination of the High Place Phenomenon," *Journal of Affective Disorders* 136: 3 (2012), 1114–20. A short commentary on the study intended for lay audiences is available here: http://bodyodd.nbcnews.com/_news/2012/03/13/10657767-that-weird-urge-to-jump-off-a-bridge-explained?lite

26 Cited in The Occulture (David Cecchetto, Marc Couroux, Ted Hiebert, and Eldritch Priest), *Ludic Dreaming: How To Listen Away From Contemporary Technoculture* (USA: Bloomsbury Academic, 2017).

9 Belly Brain
An Instrument of Ecological Perception

Doug Jarvis

I have recently been producing a series of works (under the collective title of *LIVE FEED*) that explore the concept of gut re-actions. In these works, I videotape myself eating lunch in a variety of restaurants with the goal of creating mediated episodes of engagement with my digestive system. When exhibited, selected clips are displayed on monitors and deposited around a structure lying on the floor of the gallery. The structure as a whole reflects an anthropomorphic display of the digestive system.

In her chapter "Ecologicity, Vision and the Neurological System" in the book *Art in the Anthropocene: Encounters Among Aesthetics, Politics, Environments and Epistemologies*, Amanda Boetzkes explores the effect of the Anthropocene, a new geological epoch, and its influence on the understanding of human vision. She asserts:

> If this new epoch calls us to imagine the impact of modern human life beyond the parameters of individual phenomenology, to account for systemic activity on micro and macro scales, and in relation to geological time, then there is equally a desire on the part of artists to redefine the limits of vision in order to incorporate and represent new orientations.[1]

Boetzkes's statement suggests that the existence of the Anthropocene—as a new geological epoch—is not something that is simply represented in art but rather a new environmental context that exceeds any possible mode of representation. As a result, the challenge of art, and of being an artist, shifts away from representation and towards *intervention* as a way to reveal (and engage with) how the Anthropocene is changing human perception and relations to the world. In this way art "consolidates a cultural orientation—a way of seeing."[2] Art provides possibilities for interaction that were not necessarily as visible before.

Boetzkes's proposal and her speculation on the role of artists within the Anthropocene resonates with the artistic research that I have been conducting over the last few years. During this time, I have been developing a relationship with my second brain, or what I have taken to calling my "belly brain." In his book, *The Second Brain*, Michael D. Gershon provides a detailed portrayal of the existence and function of a concentration of neurons that live in our digestive system.[3] In an attempt to render a biological framework for better understanding the cause of gastro-intestinal function and related conditions, Gershon's book provides a quantifiable portrait of what he calls the second brain. In fact, the second brain is composed of more individual neurons than those that make up the spinal cord, with a volume that is equivalent to the size of a domesticated cat brain. It produces a substantial concentration of neural activity that occurs in a place we are not accustomed to acknowledging as a site of conscientious behavior.

Doug Jarvis, *LIVE FEED*, 2013. Video still.

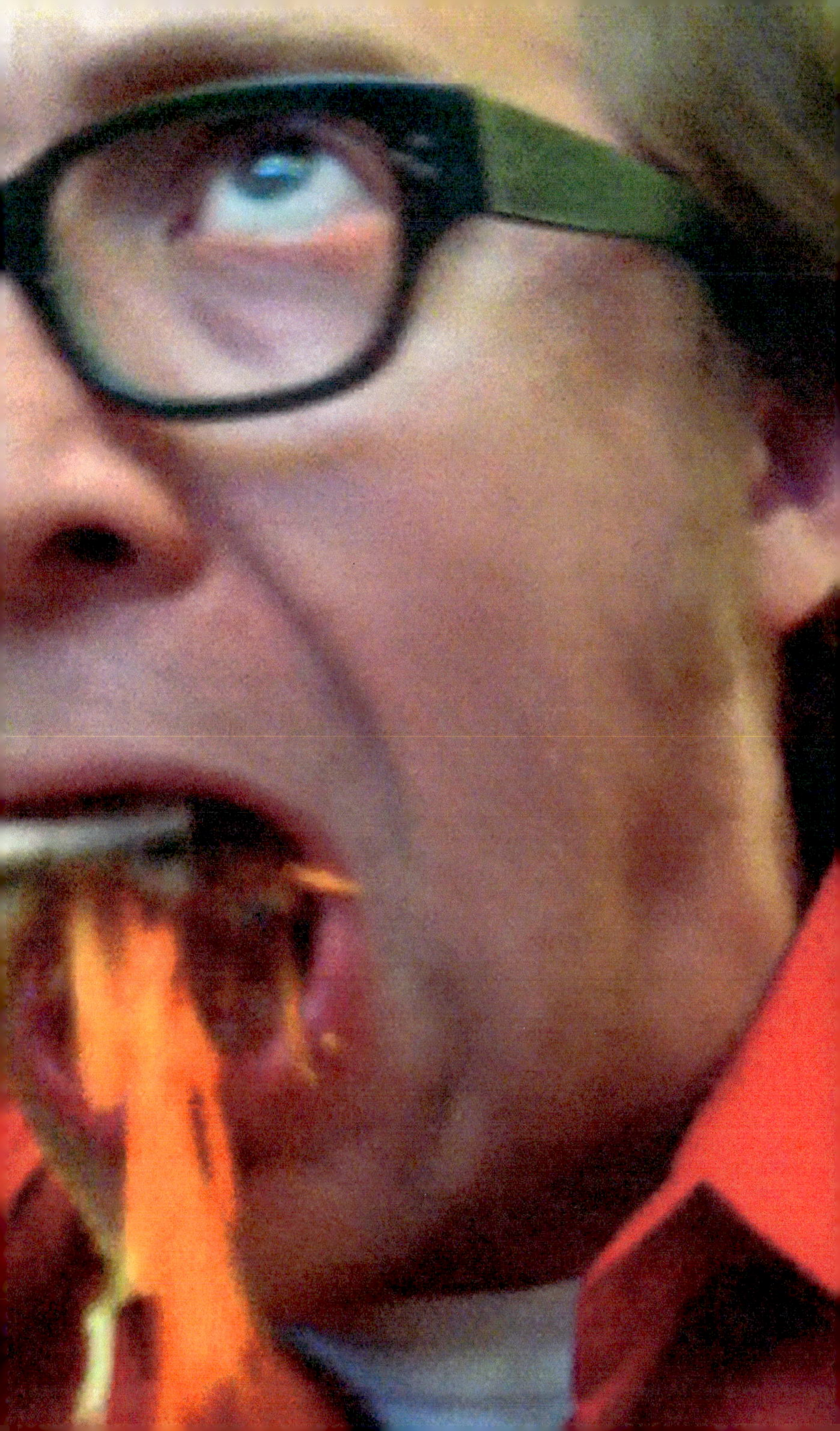

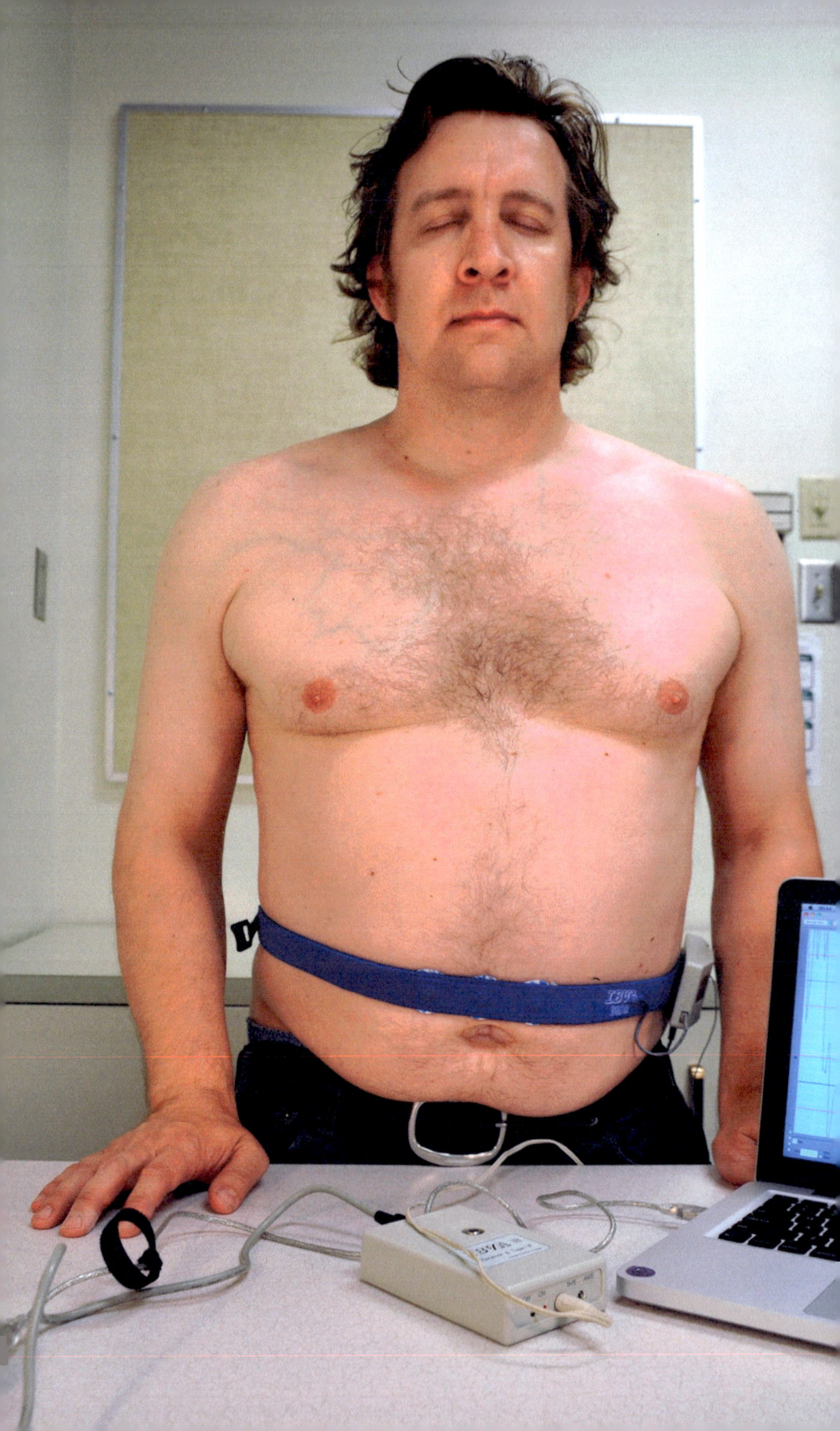

Inspired by the potential of another concentration of neurons living within my body, which apparently have the capacity to communicate (I take neural activity and communicative capacity as potentially analogous), I have taken a keen interest in trying to develop the agency of this second brain. I realize that it operates inside of my body on an ongoing basis, day-in, day-out, working in conjunction with my other life-sustaining systems as it has for over 47 years. However, now that I am more aware of its existence and how it functions as my enteric nervous system, I have taken it upon myself to pursue a stronger relationship with it. Through a series of performance actions I have compared my belly brain's brainwave activity to that of my head brain; meditated to exercise its mind; conducted a series of physical prompts to my gut region in hopes of manifesting a recognizable response. My belly has been the subject of another artist's voodoo-style psychic disturbance, and I've eaten a variety of foods to massage an anthropomorphic structure for the belly brain to manifest. Although my actions have been pretty much uni-directional, I have at least managed to influence a mild gut pang whenever I have engaged in a discussion about my belly brain and its ability to communicate. I think it is listening.

What interests me about this line of artistic inquiry in the context of Boetzkes's discussion of the Anthropocene is the potential of the belly brain to process information and to provide a point of reference for understanding how the things that we eat change our relationship to the world. How

Doug Jarvis, *Minding the Belly Brain*, 2011.
Performance photograph.

can an additional neurological processing system already active in human biology be better understood and possibly augmented to provide perceptual awareness beyond the standard perceptual knowledge we gain through the known senses? What possibilities exist for the future of being human (or for thinking beyond the possibilities of head-centered human consciousness), within a world in which our collective actions are altering geological evolution itself, if we create another mode of perception in collaboration with our second brains?

Boetzkes's speculation on the role that artists play in activating the complex terrain of human perceptual possibility has inspired me to think more deeply about the ways I might activate the belly brain. In particular, I wonder if there is, in fact, some value in developing another neural-networked system within our own agential bodies. If so, how could this network be interfaced? Can we help to encourage the development of *its* own agency, not limited to our capacity to perceive? Could this assist further engagement with our own sense of presence in the world? And, if one of the jobs of artists is to challenge the limits of these kinds of speculative relationships, then how far might these conjectures be pushed?

I think they can, at least, be pushed well beyond Gershon's initial point of reference. Gershon wrote the *Second Brain* in 1998 to provide a physiological account of the neurons in our digestive tracts and to lay the groundwork for further research into the role that our digestive system plays in the movement of food, nutrient processing, and its potential to affect our emotions. However, he has also stated that "although its influence is far-reaching, the second brain is not the seat of any conscious thoughts or decision-making .... The second brain doesn't help with the great thought processes ... religion, philosophy and poetry is left to the brain in the head."[4] This suggests that from his perspective the research does not support the potential for the second brain to be another form of consciousness. Following Boetzkes's speculative challenge however, my approach is somewhat different. Gershon can do his work, and I will do mine. My intent is not to prove that my belly brain is engaging in aesthetic word play, or even passively making decisions about whether or not it enjoyed last night's dinner. My interest is in the possibilities that arise when we spend conscious time with this concentration of neurons, in a creative way, to explore the potential of its participation in the discussion. Speculative—even rhetorical—questions are useful: Forget the question of whether it is conscious and ask the question of whether the belly brain even wants to have agency? We are told that it is similar in size to a cat brain, so perhaps that gives some imaginative indication of its potential capacity for awareness? We assume it would want to communicate, but maybe it already has something better to do and the attempt to bring it online is in vain, just another example of projecting human biases onto an unwitting companion. For all I know, my belly brain could be broadcasting a reality show about living in my physiological neighborhood and I'm just not aware which channel it's on or the device I need to watch it. In some ways it makes me think that my process of investigation is not unlike that of the SETI (Search for Extraterrestrial Intelligence) program and their persistence in scanning the known universe for signs of extraterrestrial life. The difference is that I am looking inward: my belly is my belly, but it is also somehow foreign to me—an alien on the inside.

Making a Medium Out of Non-Material Entities

In response to the *Occupy Wall Street* movement in 2011 a group of avatar performance artists from around the world gathered in Blue Mars Lite, an online environment that uses Google maps as its context. Clustering around the bronze bull located on (virtual) Wall Street our avatars engaged in a series of collective actions inspired by the real-life participants in Zuccotti Park. It turns out that Blue Mars Lite was already occupied by those that Google captured first, frozen in time by the street view image, and the performance became a show of solidarity with these ghostly inhabitants of the digital geography.

For some time now, I have been conducting research into ghosts and other non-material entities as an audience for my work. It is a prospect that can provoke vertigo. It has also provided an interesting challenge to think about the ways that exploring the agency of a non-material entity might alter aspects of the world that I have grown to know. The thought of my Facebook profile picture, for example, having what might be referred to as a mind of its own pushes the parameters of the sensible world and destabilizes the ground upon which I base my reality. I feel similarly about the idea that a ghostly human presence might have the ability to exercise energy in a way that could influence the space around me, based on its own choosing. The idea is unsettling, but also exciting. The potential of non-material entities to act of their own accord would expand the understanding of how the world is composed, distorting many values that I have grown to believe. Part of the work that I have taken on is to propose questions that will disturb deep-seated perceptions and contest the parameters of how participation in the world starts and stops. What would it mean for an avatar or a ghost to have agency, in human form or otherwise? What would change if we accept that they may be present in some form and that when in this form they are able to communicate with each other (and maybe with us as well)?

Doug Jarvis, *Virtual Occupy Wall Street*, 2012.
Documentation of avatar action in Blue Mars Lite.

These questions have proven to be good exercises. They help to reveal my biases and the assumptions that I make about how I am physically and emotionally composed. The act of speculating, when taken seriously, unearths prejudices and openly challenges the patterns of thought and the constructs of reality. It inspires me to practice different ways of existing in the studio of day-to-day existence and to really contemplate the conditions of my work. It makes me question what I make, the medium that I use, and whether or not it could be perceivable to something like a non-material entity. These are works that use my physical and psychological systems, agents, and processes as the medium. They introduce questions about what it's like to be a human in this contemporary age. They gesture towards the limits of the human as well. I explore states of being and non-being. I investigate the possibilities of consciousness and non-consciousness. The works I make propose opportunities to engage with other entities that may or may not be able to reciprocate, pushing the boundaries of what can be considered collaboration. Is it collaboration if the entity that you are working with might not have the capacity to participate in a way that you are able to recognize? Is this just personal delusion, or is it reaching for a possibility that just needs a bit more time, a little more fuel, another inch of tether to stretch to make a connection? Is this the kind of action that can be shared by others to help bring it into the domain of contemporary art? Or, is it just wading around in uncertainty and speculation with too many open variables that will only continue to drift off into the ether without any way to bond with other agents into a self-aware form?

These are abstract questions, but by asking them I have developed a framework for a potential relationship between avatars, ghosts, and my belly brain. Derived from the intersection of my non-material output (online activity, theoretical speculation) and my real-life collaborations with other humans, I have formulated a way of imagining how I might be able to bridge communication between the various non-material entities in my life. I consider ghosts and avatars as the past and future residue of human participation in this world—and I consider my belly brain as a conduit for engaging with them. Considering that the belly brain is composed of biological neurons that have the capacity to communicate using electro-chemical systems, in theory, it should be possible to interface with the energy systems employed by avatars and ghosts. What would my belly brain need to do to be open to this possibility? What would I need to do to encourage/foster/coax it to exercise its material and non-material aspects so that it can be an instrument to communicate with future and past energy residues, with avatars and ghosts?

To support the development of this framework, and to help encourage my belly brain to participate in this mediation program, I have engaged in a few exercises to help connect with my non-material friends. I have taken it upon myself to foster a supportive context for the development of non-material agency in the world. I have realized that my process is fueled by western social and cultural ways of thinking. The fact that I have to engage tools of imagination as a means of proposing such relationships is proof of western social and cultural bias; for example, I have not found any customs that provide the space to have a discussion between the ghost of my grandmother and my Second Life avatar *Tran Spire*. Both of these non-material entities carry nuances of my life. And, to be fair, they do live on different platforms, in spaces that do not necessarily interact. This is where the belly brain comes in. As I have stated, from what I can tell, the belly brain has a foot in both worlds. The neurons of which it is composed live in the tissues of my digestive tract and their electro-chemical communications have the potential to register on the electromagnetic spectrum. In my project *Minding the Belly Brain*—in which I connect a brainwave sensor to my belly—my goal is to encourage my belly brain to exercise its mind. My hopes are to strengthen its ability to be an intermediary and employ telepathic techniques in hopes of establishing contact with my grandmother's ghost and my Second Life avatar. In a sense this has been my motivation all along, to pursue lines of artistic research that forge a relationship with things that may or may not exist, to develop strategies and opportunities for communication with me, if they want.

I am driven by this interest to speculate. How can the belly brain provide another tool with which to explore the world as we encounter it? And what would be the limits of this tool—or the point at which it fails, breaks down, or speaks back to us in ways that are unexpected? When Boetzkes talks about the ways that technologies only reveal their true meaning when pushed to a point of failure, she is not only commenting on a philosophical situation, she is giving artists permission to challenge ourselves to come up with even more speculative attempts to knock human perception off of its familiar axis and un-root the ways in which we assume (artificially) our dominant place in nature.[5] The combination of how artists inhabit the world in which they live, for artistic ends, and the engagement of their work by viewers, audiences, and participants provides an appropriate space to look at how the line between where the artist and the world in which they work become better understood as two-sides of the same coin.

Go With the Gut: Interfacing Mutualistic Relations

To expand the scope of my belly brain research I developed a series of performance actions that took the idea of second brain beyond just a shared thought and put it into the realm of cause and effect. In a series of works titled

Doug Jarvis, *WOW! Tran Spire visits World of Warcraft*, 2007. Documention of avatar action.

Gut Re-Actions I explored the potential of the gut to direct a sequence of actions in a gallery or auditorium context. Performing in front of an audience at the University of Washington Bothell, for instance, I conducted a series of spontaneous actions on my stomach and gut area using intuition and physiological response as guides. I snapped my belly with large elastic bands, stroked my stomach and head

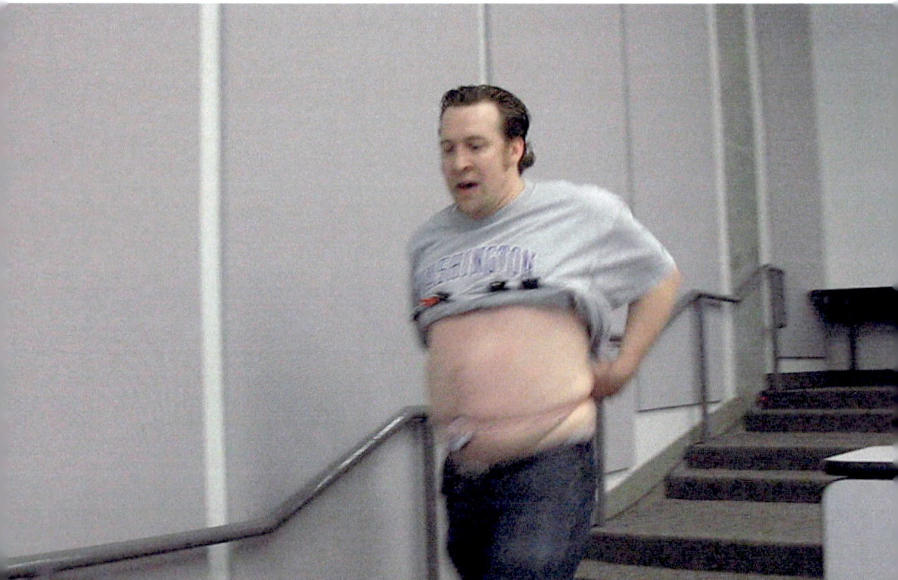

Doug Jarvis, *Gut Re-Actions*, 2013.
Documentation of a performance at UW Bothell.

simultaneously with thin-wire head massagers (like the ones you can buy at a kiosk at the mall), chugged cans of energy drink, handed out potato chips to anxious onlookers, and ran around the perimeter of the theater, up and down the stairs, wrapping my torso in butcher's twine to hold in place a UW lollipop that I had affixed to my belly button. The hour-long sequence folded together seemingly random acts into a curious display of non-conscious activity. What held the actions together was the gut feeling that they all had something in common.

The *Gut Re-Actions* series was necessary as a way to physically engage with the belly brain and to create a parallel discursive space alongside the conceptual, non-material aspects of the work. The goal was to establish a more tangible method for making the belly brain publicly present. The result was a series of inspiring conversations that I had after each performance where members of the audience would share stories of their own digestive systems and the often-complicated relationships that they managed with its bacterial make-up. Eating became a common topic, leading to deeper contemplations about what exactly goes on inside the gut. How does it function in terms of food processing and nutrient extraction? As I started to explore the composition of the digestive system and the bacteria that live inside of the small and large intestine, I became more curious about how this micro-biome functions and the space where the neurons of the belly brain come into contact with the colonies of bacteria in the gut. New research goes as far as to suggest that the micro-biome plays a role in the control of human emotion,[6] which, of course, makes me wonder about how the belly brain interacts and collaborates with the environment around it. In the gut, bacteria work together to process the food that we eat and to influence how we feel. This seems significant and it makes me wonder what else is going on?

In particular, I want to see if there is a line of inquiry that can explore the stomach as a mutualistic companion system and see how it could be used to help render different ways of connecting to the world. If the Anthropocene is a global condition that implicates humans as the drivers of environmental change, then how can our relationship with the world as perceived by the belly brain and its companion micro-biome contribute to this discussion? How can this relationship be used as a speculative medium to propose possibilities that go beyond standard human communications to augment our ability to not just communicate within the world but to become more attuned to the impact we are having as well?

One way to do this is simply to look for opportunities to behave differently, giving agency to a different form of being in the world, whether inspired by gut feelings or not. Since 2012, I have been doing performance actions as part of the Open Actions performance group in Victoria. Each month we meet at a predetermined location, date and time, and begin the performance with a series of independent actions. Sometimes individual performances turn into collective forms, sometimes they just evolve on their own, without any necessarily prescribed outcomes. It is an effective way to be present in a public space, protected by the presence of a group, yet open to whatever sort of action or gesture spontaneously emerges. Key to the philosophy of the group is the sensibility of being open to what will come about, to what wills itself to manifest in some form, and to what then becomes the generator of the performance. The actions are an attempt to access the terrain of the unconscious and of non-conscious perception. Or, perhaps, these actions provide a context where the non-conscious aspects of mutualistic relations can come together and make something present. By working as a group the events take advantage of the moment in an attempt to be collectively present.

During an Open Action event, it is not uncommon for one artist's action to suggest parameters for a similar style of gesture, or sound, from another artist— for example, one person timing their footsteps to the cadence of another person shouting. Or, one artist covers his or her head, and another follows, blind to vision, but active in the soundscape that emerges as their collective central focus. These are not really coincidences as much as examples of artists attuning to the circumstances of the site in which we perform and to each other as contributing factors to the overall aesthetics of the event. The communications bring something forth in a variety of dimensions and manifest as a series of gestures or movements, a sequence of actions that reveal a non-narrative that the non-conscious part of us is somehow able to conceive. Maybe the non-conscious is taking advantage of the consciousness's lack of head-brain dominated intentionality? Perhaps this is one possible forum in which a new matrix of human/non-human and conscious/non-conscious systems can establish the possibilities of their own kind of public engagement, their own open actions? It's very simple in some ways—artists responding spontaneously to other artists—but the mechanism that governs this kind of response is not easily understood intellectually. It requires gut communication.

In this way, the belly brain is both a metaphor for these kinds of non-conscious systems and is itself a real phenomenon from which we have much to learn. I believe that building a relationship with this collective seat of perceptive and emotional wisdom can allow us to engage the world in a deeper way. I think that exercising a program of material and non-material body awareness can augment how we acknowledge our presence as active human agents on the planet. I am convinced that there is much to learn by fostering a relationship with the millions of bacterial cell structures that eat our food and influence our moods. The question is what do we do now and how do we actively cultivate these sorts of mindsets and relationships?

My suggestion is that we follow. We listen. We give the world as it still is a chance to grow again. If we can change our non-renewable resource extraction processes, we can figure out how to become a part of the world again, working with the earth as opposed to against it. If we could somehow figure out how to eat plastic as a way to magically compost it, then we would be getting somewhere. The human-centered ego will not easily lose its will to persist. This is why we need to figure out how to use everything that our guts have to offer. This is where we check our species privilege and use the sustainable tools we are creating to collaborate with our intuition and our diligent cultural colleagues who know how to speak with the land, with the water and the air. We can learn how to listen to our guts and to hear what they are hearing on the cosmic airwaves of the micro-organic. We might wonder how we can give space to the power of entities that we don't even know how to perceive, but we are the humans and we are the ones responsible for the emergence of an anthropocentric perspective. We also need to be the ones asking the questions and wondering what we can do about it. It's time to make our shit and listen to it too.

When I Move You Move
Inter-Entity Mirror Neuronics

Moving forward, I am developing another series of actions to further engage the question of how we can become more of the world with the tools and processes that we have developed within the world. Now, for clarity, this is my head brain talking. I am familiar with being the subject of questions, curiosities, pokes and prods, in hopes of better understanding what I am made of and how I function. I am a wonder to behold; and the head brains of many other humans have paved the information highway with the magic that we bestow and the potential folded up in our cranial pathways. Our head brains are used to certain logic systems and have become accustomed to being the primary cognitive tools that humans have at our disposal. Therefore, I am going to initiate a series of events that will explore the potential for my belly brain to exhibit traits found in my head brain, specifically, its ability to mirror other entities. This is probably just a tactic to distract the head brain from thinking that the belly brain can only function on its terms, that a brain is a brain only as we know it.

There is a theory, derived from research on primates, that motor command neurons in the brain are activated when they witness a specific action performed by another agent.[7] This ability for parts of the brain to mirror actions that it sees is thought to aid in our ability to learn, be empathetic, and be introspective.[8] The sensation that you have to yawn when you see someone else yawn is an example of mirror neurons at work. Since the neurons in the belly brain are already understood as responsive to their digestive environment (when I move, you move) I wonder if they are

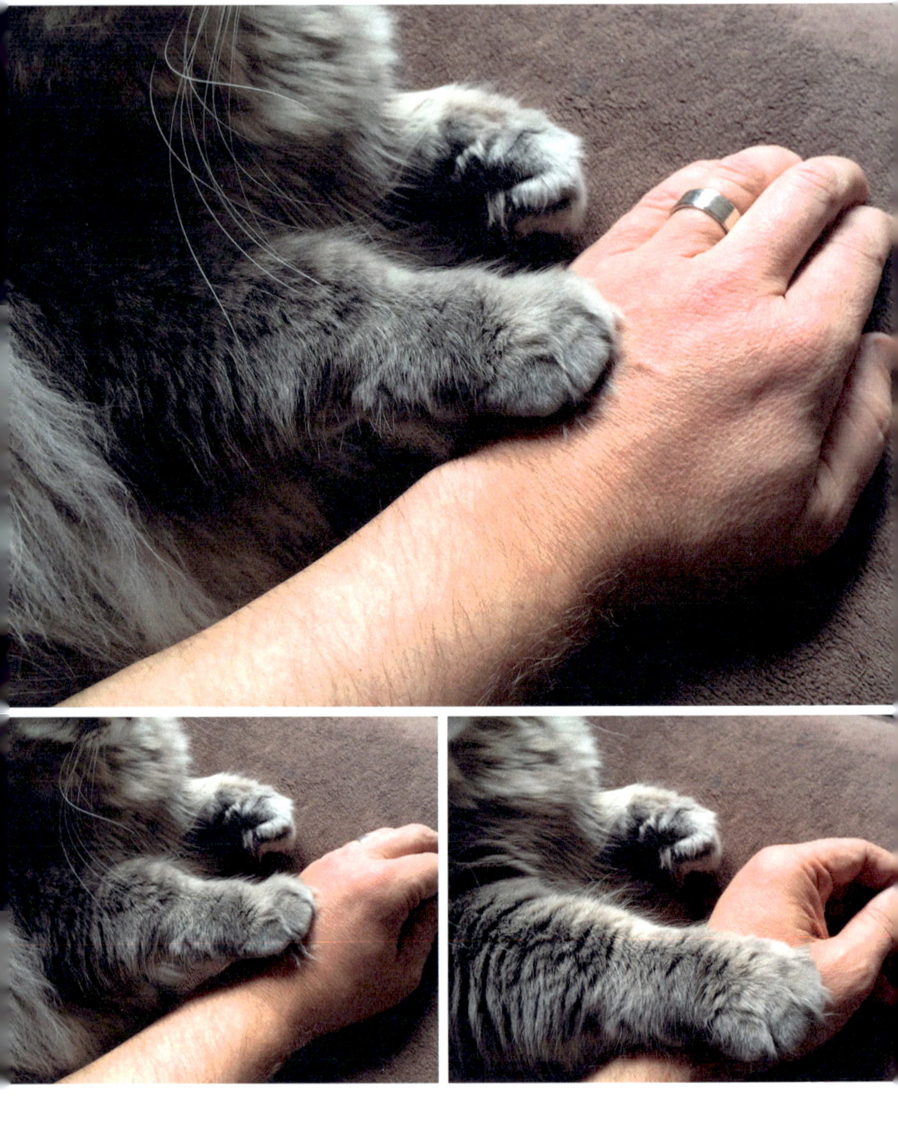

Doug Jarvis, *Paws: Exercise in inter-entity mirror neuronics*, 2016. Photographs.

also able to respond to similar processes in other entities? Or other animals, let's say, like my house cat, Ava. When Ava responds to an action of mine, is she simply witnessing me visually or is her belly brain responding somehow to what I have done? And how does my belly brain register this action in return? How might it respond in the way that a motor command neuron in the head brain might respond? Does it fire in recognition, and if so, to what, or to whom would it be signaling? My plan is to exercise a series of actions that mirror other entities, and then to observe the ability of the neurons in my gut to manifest presence by reflecting other entities. Will Ava be able to communicate with my belly brain? I realize that the relationship between my cat and my gut brain is merely a coincidence, but I wonder if there is more to this reference than just a shared volume of neurons?

By introducing companion species into my framework with ghosts and avatars, will the belly brain find a more sympathetic circumstance in which to engage in mirror neuronics, in inter-entity mirror neuronics? How will the belly brain render instances of being like the world around it? Around us? How will I know? When will I know? Is this collaboration or am I just going to, once again, subject my belly brain to a series of experiments to see if it responds in any way to a cat, a thing that it shares a characteristic with? Am I going to need to see how it is similar in order for me to stop projecting unusual circumstances on it just to satisfy a curiosity to see what will happen? What do I hope to learn from this line of inquiry, from this line of testing? Is the belly brain becoming a vicarious outlet for me? Am I letting it be what it can be? The fact that I am asking such questions is starting to touch a nerve.

Is this then not the space that Boetzkes is creating for artists in relation to an ecological consciousness? Humans need to acknowledge some kind of responsibility for our recent history of non-sustainable practices and our ignorance of the reciprocal nature of ecological systems. Even the cynic who does not want to pay for the loss of nature due to industrial energy extraction still has to invest in some kind of technology to accommodate for the change in environment. The fact that we are even discussing how things *have* changed means that we are in some way implicit to the current condition. The changes that humans have generated on a geological evolutionary tract are changing the world as we know it, and in turn changing the way that we are able to perceive it too. Isn't this what Boetzkes is talking about, that artists are the ones who will present the options for consideration? Not just as representations of their artistic gestures and constructs, but as the litmus test strips of how human imagination is stuck in its own cycle of reflection and needs some help from other logics to inform a way forward?

Are inter-entity mirror neuronics and the belly brain then not viable options to try and develop non human-centric ways of engaging with (and within) the environment? Are they not ways to collaborate with our non-human and non-material companions? As Boetzkes says, "It is therefore worth examining the diverse approaches that orient art toward an ecological consciousness, as well as the political, social, and aesthetic issues that these new forms address."[9] I take this to mean that without a concerted attempt to sacrifice the boundaries of the human we can never hope to actually participate in a properly environmental discourse. And without a properly environmental discourse, we can never hope to actually be a healthy and collaborative part of the ecology of the world that we inhabit. In this spirit, I encourage you all to join me in creating tools of engagement by performing actions that don't just demonstrate what the world is already like, but that change the way that we are of the world by mirroring the world that makes us who we are. What does your belly brain think about that?

Notes

1 Amanda Boetzkes, "Ecology, Vision and the Neurological System" in *Art in the Anthropocene: Encounters Among Aesthetics, Politics, Environments and Epistemologies*, eds. Heather Davis and Etienne Turpin (London: Open Humanities Press, 2015), 272.

2 Ibid.

3 Michael D. Gershon, *The Second Brain: A Groundbreaking New Understanding of the Nervous Disorders of the Stomach and Intestine* (New York: Harper Collins Publisher, 1998), xiii.

4 Michael Gershon, cited in Adam Hadhazy, "Think Twice: How the Gut's 'Second Brain' Influences Mood and Well-Being." *Scientific American*, February 12, 2010. http://www.scientificamerican.com/article/gut-second-brain/

5 Amanda Boetzkes, *The Ethics of Earth Art* (Minneapolis: University of Minnesota Press, 2010), 4.

6 Dan Hurley, "Your Backup Brain," *Psychology Today,* November 1, 2011. https://www.psychologytoday.com/articles/201111/your-backup-brain

7 Giacomo Rizzolatti and Luciano Fadiga, "Resonance Behaviors and Mirror Neurons," *Italiennes de Biologie* 137 (1999): 85–100.

8 Giacomo Rizzolatti, cited in Vilayanur Ramachandran, "The neurons that shaped civilization," *TEDIndia*, November 2009. https://www.ted.com/talks/vs_ramachandran_the_neurons_that_shaped_civilization

9 Boetzkes, *The Ethics of Earth Art,* 25-26.

Contributors

Shannon Bell is a Professor of Political Science at York University (Toronto). Her books include: *Fast Feminism* (2010), *Reading, Writing and Rewriting the Prostitute Body* (1994); *Bad Attitude/s on Trial* co-authored (1997); *The Book of Radical General Semantics* (2016) *Subversive Itinerary: The Thought of Gad Horowitz*, co-edited (2013) and *New Socialisms* co-edited (2004). Bell is currently working on *Shooting Theory*—video-imaging philosophical concept such as Heidegger's "stillness," Husserl's "epoché," Batialle's "waste," Weil's "attention," Deleuze's "deterritorialization," Virilio's "vision machine," Levinas's "elemental" and Mallin's "sinuosity." Her vimeo.com name is FastBodies.

David Cecchetto is Associate Professor in the Department of Humanities at York University (Toronto). He is the author of *Humanesis: Sound and Technological Posthumanism* (University of Minnesota Press, 2013) and co-author (with fellow members of the experimental theory group "The Occulture") of *Ludic Dreaming: How To Listen Away from Contemporary Technoculture* (Bloomsbury, 2017). As a quasi-practicing non-musician, David has presented work in Canada, Mexico, Russia, the UK, and the USA.

Mark A. Cheetham has published extensively on historiography and art theory, abstract art, and art in Canada. A Guggenheim Fellow, he has also received writing and curatorial awards from the College Art Association of America and the Ontario Association of Art Galleries.

His book *Landscape into Eco Art: Articulations of Nature since the '60s* will appear with Penn State UP in 2017. He is a professor of Art History at the University of Toronto.

Ted Hiebert is Associate Professor of Interdisciplinary Art at the University of Washington Bothell. He is the author of *In Praise of Nonsense: Aesthetics, Uncertainty and Postmodern Identity* (McGill-Queens University Press, 2012), *A formalized forum for informal inquiry* (Noxious Sector Press, 2015), and (with David Cecchetto, Marc Couroux and Eldritch Priest) *Ludic Dreaming: How to Listen Away from Contemporary Technoculture* (Bloomsbury, 2017).

Doug Jarvis is an artist, curator, administrator, and educator living in Victoria, BC. He is a founding member of the avatar performance art group Second Front and the Noxious Sector Arts Collective. His individual and collective works have been presented at artist-run centres, galleries, museums and festivals across Canada, the United States, Europe and Asia. Jarvis is an active member of the Victoria arts community and sits on a variety of municipal committees, non-profit, and artist-run centre boards.

Serena Kataoka is an arts administrator, artist, and academic based in North Bay ON. By day she is the Executive Director of the White Water Gallery, and by night she teaches social justice oriented courses at Nipissing University. Her research on urban theory has been published in international venues, but her artistic practice has tended to be site-specific. Threading through all of her work is a commitment to improving our ways of living together *here*.

Anne Pasek is a doctoral student at New York University's Department of Media, Culture, and Communication. Her dissertation examines the history, visual culture, and politics of geoengineering and other speculative climate technologies. She holds previous degrees in art history from McGill University and the University of Alberta.

Andrew Pendakis is Assistant Professor of Theory and Rhetoric at Brock University and Research Fellow at Shanghai University of Finance and Economics. His interests lie in critical theory and political philosophy with a specific focus on the history, theory and practice of centrism. He has published essays on thinkers such as Heidegger, Badiou, Deleuze, Blanqui and Adorno and on topics ranging from the nature of dialectics to the phenomenon of plastic.

Maria Whiteman's art practice explores relationships between industry, natural energy, and the place of animals in our cultural imaginary. She has exhibited her work widely at venues such as the Caetani Centre (Vernon, BC), the Banff Centre for the Arts, Urban Video Project (Syracuse, NY), The Museum of Walking (Tempe, AZ) and Latitude 53 (Edmonton, AB). Her critical writing has been published in *Public: Art/Culture/Ideas, The Minnesota Review, Antennae* and the *John Hopkins Guide to Literary Theory and Criticism.*

Made in the USA
Las Vegas, NV
10 November 2026

57858721R00119